5-HTP

Nature's Serotonin Solution

5-HTP

Nature's Serotonin Solution

RAY SAHELIAN, M.D.

AVERY PUBLISHING GROUP

Garden City Park • New York

The medical and health procedures contained in this book are based upon the training, research, and personal experiences of the authors. Because each person and situation is unique, the author and the publisher urge the reader to check with a qualified health professional when there is any question regarding the presence or treatment of any abnormal health condition.

The publisher does not advocate the use of any particular diet or health program, but believes that the information in this book should be available to the public.

Because there is always some risk involved, the publisher and author are not responsible for any adverse effects or consequences resulting from the use of any of the suggestions, preparations, or procedures discussed in this book. This book is as timely and accurate as its publisher and author can make it; nevertheless, they disclaim all liability and cannot be held responsible for any problems that may arise from its use. Please do not use the book if you are unwilling to assume the risk. Feel free to consult with a physician or other qualified health professional. It is a sign of wisdom, not cowardice, to seek a second or third opinion.

Cover design: Doug Brooks
Typesetter: William Gonzalez
In-house editor: Lisa James

Avery Publishing Group, Inc.
120 Old Broadway
Garden City Park, NY 11040
1-800-548-5757

ISBN: 0-89529-903-8

Printed in the United States of America

10   9   8   7   6   5   4   3   2   1

CONTENTS

Acknowledgments

Stan Bazilian, M.D., of Philadelphia, Pennsylvania, is a psychiatrist who uses nutritional therapies in his practice along with pharmaceutical medicines.

Susan Busse, M.D., of Palatine, Illinois, practices nutritional medicine.

William F. Byerley, M.D., is a member of the Department of Psychiatry, University of Utah, in Salt Lake City.

Barry Elson, M.D., Medical Director of Northampton Wellness Associates in Northampton, Massachusetts, specializes in nutritional and environmental medicine.

Dale Guyer, M.D., Director of the Center for Innovative Medicine, in Indianapolis, Indiana, focuses on nutritional therapies and antiaging medicine.

Peter Hauri, Ph.D., is Director of the Mayo Clinic Sleep Disorder Laboratory in Rochester, Minnesota.

Joseph P. Horrigan, M.D., board certified in Child Psychiatry, is an Assistant Professor at the University of North Carolina, Department of Psychiatry, Developmental Neuropharmacology Clinic, in Chapel Hill.

Jeanne Hubbuch, M.D., of Newton, Massachusetts, integrates traditional and alternative medicine in her private practice.

William Kracht, D.O., FAAFP, of Quakertown, Pennsylvania, is in private practice.

Richard Kunin, M.D., of San Francisco, California, is President of the Society for Orthomolecular Health-Medicine.

Jay L. Lombard, D.O., a Diplomate in Neurology of the American Board of Psychiatry and Neurology, is Clinical Assistant Professor of Neurology at Cornell University in Ithaca, New York. He combines traditional and nutritional therapies to treat neurological disorders, and is coauthor of *The Brain Wellness Plan*.

Patrick Melese, D.V.M., of San Diego, California, is a veterinarian who combines traditional medicine and nutritional therapies in treating behavioral problems in animals.

Larry Pastor, M.D., Clinical Assistant Professor of Psychiatry and Human Behavior at George Washington University Medical Center in Washington, D.C., is an expert in the psychopharmacologic therapy of anxiety and depressive disorders.

Bob Perry, Ph.D., completed his doctoral thesis at the University of California—Santa Barbara on the metabolism of tryptophan to melatonin. He is currently a nutrition advisor to Green Foods in Oxnard, California.

Ascanio Polimeni, M.D., is director of the PMS and Menopause Center in Rome and Milan, Italy.

Christian Renna, D.O., maintains private practices in Dallas, Texas, and Santa Monica, California, and specializes in preventive and nutritional medicine.

Douwe Rienstra, M.D., of Port Townsend, Washington, is in private practice.

Priscilla Slagle, M.D., is a psychiatrist who maintains private practices in both Encino and Palm Springs, California. She is the author of *The Way Up From Down*. Her website is www.thewayup.com.

Karlis Ullis, M.D., is Clinical Professor at the University of California—Los Angeles, and maintains a private practice in Santa Monica, California. Dr. Ullis specializes in nutritional therapies, sexual medicine, and antiaging medicine.

Nathan D. Zasler, M.D., is editor-in-chief of *Neuro-Rehabilitation: An Interdisciplinary Journal*, and is medical director of the Concussion Care Centre of Virginia, near Richmond. He uses 5-HTP in certain patients with brain injuries.

I would also like to thank the following people for their information and insight: David Blanco; Will Block; Heinrich Kaden; John Morgenthaler; Mark Olsen, M.S.; Sid Shastri; David Stouder; Blake Smith, R.Ph.; Ronald G. Sturtz; and Sunny Yiu, Pharm. D.

I would also like to thank Barbara Hirsch, publisher of the newsletter *Obesity Meds and Research News*. Her website is www.obesity-news.com.

INTRODUCTION

Prozac. Zoloft. Paxil. What do all these popular anti-depressant medicines have in common? They all help improve mood by increasing brain levels of an important chemical known as serotonin. Scientists believe serotonin to be an important factor in many disorders, ranging from depression and anxiety to obesity and premenstrual syndrome.

If the problem is a shortage of serotonin, why not just take serotonin itself? The answer is simple. If you take serotonin directly, it cannot cross from your bloodstream into your brain. Is there a natural way to raise serotonin levels? Yes, there is.

The supplement 5-hydroxytryptophan (5-HTP) is a substance your body creates from an amino acid called tryptophan. 5-HTP, in turn, is used to create serotonin. Unlike serotonin itself, 5-HTP can readily make its way from your bloodstream to your brain, where it can quickly be converted into serotonin.

Does this mean 5-HTP is as effective, and works exactly the same way, as Prozac and the other antidepressants? No, but there are enough similarities in their ultimate purposes for doctors to show great interest in 5-HTP as not only an antidepressant, but also as a weight-loss therapy, antianxiety agent, sleep aid, and more. The 5-HTP and serotonin stories are complex, yet fascinating. I have gathered information from decades of studies, interviews with 5-HTP users, reports from my patients, my personal experiences, and interviews with doctors across the country who use 5-HTP in treating their patients.

The true promise of 5-HTP will most likely come from its intelligent use in combination with a number of other medicines, nutrients, amino acids, herbs, and hormones. The process of combining natural supplements and medicines is an area that has been little explored. As we move into the twenty-first century, we may just find that many of the solutions to a number of difficult medical and psychiatric conditions were at our fingertips all along. We just did not know how to best use what we already had.

After answering some commonly asked questions about this intriguing new supplement, I will tell you about my own experiences with 5-HTP. I will then explain the connection between 5-HTP and serotonin before addressing several disorders, both psychological and physical, for which 5-HTP shows promise. Finally, after telling you what the experts have to say, I will give you important information on how to use 5-HTP, including advice on dosages, complementary therapies, cautions, and side effects.

In presenting the leading-edge research that's been done on 5-HTP, I wish to make it clear that our understanding of 5-HTP's clinical role and safety is still in its infancy. I have written this book to provide balanced information on a supplement that is already on the over-the-counter market. I do not sell 5-HTP or any other supplement, nor do I endorse any products.

So keep in mind that while the research on 5-HTP is exciting, it is also still ongoing. In any situation in which you are not sure about the appropriate use of a supplement or other therapy, see your health care provider. For the latest updates, see www.raysahelian.com.

I hope *5-HTP: Nature's Serotonin Solution* will make both the medical community and the general public more aware of some of the benefits 5-HTP can provide. This may, in turn, lead to further research into this long-neglected nutrient.

CHAPTER 1

QUESTIONS FREQUENTLY ASKED ABOUT 5-HTP

Dozens of new supplements are introduced each year in the health supplement industry. The majority of these are like shooting stars that fade away when consumers realize that they just do not work as promised. 5-HTP will not be one of those shooting stars. With time, it will become a more popular supplement.

What is 5-HTP?

5-HTP is short for 5-hydroxytryptophan. This molecule is made in the body from the amino acid tryptophan, and is used to create serotonin, an important chemical found throughout the body, particularly in the brain, digestive system, and blood cells. Tryptophan is one of the essential amino acids, which are amino acids the body cannot create. Tryptophan must be obtained through the diet, from foods containing protein such as meat, fish, dairy products, and certain legumes.

Where does the 5-HTP I buy come from?

The 5-HTP sold in stores comes from the Griffonia seed,

which in turn comes from a shrubby African tree grown mostly in Ghana and the Ivory Coast. Certain African tribes crush these seeds, which are the size of small peas, and give them to children who are in need of calming. In the early 1980s, when the patents on Valium and Librium ran out, European researchers started looking for alternative anti-anxiety agents. Their search led them to 5-HTP, and to the Griffonia seed.

There are at least three European pharmaceutical companies that extract this nutrient from these seeds, which contain from 5 to 10 percent 5-HTP (Fellows, 1970). This pharmaceutical-grade material is then imported by vitamin companies in the United States for encapsulation and branding.

5-HTP can also be made synthetically in the laboratory, but this process, at present, is more expensive than extracting it from the Griffonia seed.

When did 5-HTP become available?

This supplement was first introduced to the American over-the-counter market in 1994. Initially, only a couple of small companies sold it. However, there are now quite a number of small and large companies marketing this nutrient. Since 5-HTP cannot be patented, any company can manufacture and sell it. This supplement has also been available for a number of years through compounding pharmacies, or pharmacies that prepare specific medicines and combinations not normally available in regular pharmacies.

You say that 5-HTP is created from tryptophan. Wasn't tryptophan banned?

While 5-HTP is currently available without a prescription, tryptophan was banned by the Food and Drug Administration (FDA) in 1989. This happened after physicians in

New Mexico noted three patients with an unusual disease marked by severe muscle pain and a high number of eosinophils, white blood cells involved in allergic reactions, in the blood (Belongia, 1996). Subsequently, more cases— eventually, up to 1,500—were discovered across the country. It was determined that the individuals who had developed this disease, called eosinophilia-myalgia syndrome (EMS), had been taking a tryptophan product made by a Japanese company called Showa Denko. This company was using a new manufacturing process that allowed trace contaminants to be mixed with the final product. No tryptophan from other manufacturers was found to cause EMS. Nevertheless, the FDA removed all tryptophan products from the market (Shapiro, 1996).

Did a contaminant really cause EMS, or was tryptophan the culprit, as some supplement-industry critics have charged? An editorial in the highly respected *Journal of Rheumatology* is quite convincing. Edward Belongia, M.D., from the Marshfield Medical Research Foundation in Marshfield, Wisconsin, and Gerald Gleich, M.D., from the Mayo Clinic and Foundation in Rochester, Minnesota, exhaustively reviewed all the available evidence. They came to this conclusion: "The evidence that a trace contaminant in Showa Denko caused the EMS epidemic is powerful and compelling." (Belongia, 1996) Even with all the "powerful and compelling" evidence pointing to tryptophan's innocence, you are still likely to encounter statements that blame tryptophan itself for the EMS problem. Moreover, this criticism may eventually be leveled at 5-HTP, especially as this nutrient's popularity grows. (Two factors that could limit 5-HTP's popularity are its cost and the fact that it takes time to learn how to use it best.)

By the way, despite the ban on over-the-counter sales, the FDA does allow tryptophan to be added to infant formulas and mixed with intravenous fluids for hospital patients. It is also available through compounding pharmacies by prescription.

So is 5-HTP safe?

As of the writing of this book, and after three decades of research with 5-HTP, no serious side effects have been reported with the appropriate use of this nutrient. The side effects that do occur, generally at higher dosages, include nausea, daytime sleepiness, and reduced sex drive. All of these effects can be reversed by discontinuing the supplement.

It helps to keep in mind that risk is always relative. We expose ourselves to risk anytime we take any medicine, even aspirin or high dosages of vitamins. As with practically every course of action we take in our daily lives, we normally weigh the potential gain against the potential harm. If a person suffers from obesity, depression, anxiety, or insomnia, or some other medical or psychological condition, there are hazards in not taking any action at all. Relationships, as well as school and work performance, could suffer, and both obesity and depression are associated with a shorter life span. Also, a person may choose to take pharmaceutical medicines, which themselves have potential side effects.

Each individual has to learn as much as he or she can about different therapeutic options, and then decide which of these options presents the most value with the least peril. I have tried to present a balanced viewpoint regarding the benefits and shortcomings of 5-HTP. After reading this book, it will be up to you and your health care provider to decide which course of action is best for you. I can say that you should use the lowest dosage of this nutrient that is effective, combine 5-HTP with other natural therapies, take occasional breaks, and limit its continuous use to less than three months. For more information, see Chapters 11 and 12.

Why isn't there more research on 5-HTP?

In 1988, William F. Byerley, M.D., and his colleagues from the University of Utah published a journal article thoroughly reviewing the studies that had been done on 5-HTP. Dr. Byerley tells me, "Our review of published studies indicated 5-HTP to have antidepressant activity. In 1989, we set up a double-blind study funded by the National Institutes of Health to examine the antidepressant qualities of 5-HTP in more detail. However, when the tryptophan scare occurred, our study was tabled. I stopped studying 5-HTP and changed my focus to a different medical area."

It is quite likely that many researchers in the United States and all around the world abandoned their interest in 5-HTP after the tryptophan episode. It is unfortunate that this one contamination incident has had such profound consequences. I hope this situation will change.

For which conditions can 5-HTP be used, and why?

Formal human research with 5-HTP has so far been limited, and most doctors have very little practical clinical experience with this nutrient. However, over the past three decades, scientists have tested 5-HTP for use in treating the following conditions:

- Obesity and overweight (Chapter 4)

- Depression (Chapter 5)

- Anxiety disorders (Chapter 6)

- Insomnia (Chapter 7)

- Fibromyalgia (Chapter 8)

- Other disorders, including premenstrual syndrome and migraine headaches (Chapter 9)

All of these conditions are related to the chemical serotonin. The concept of a "serotonin deficiency syndrome" that causes a variety of medical and psychiatric conditions is gradually taking hold. This concept was brought before the public with the popularity of Prozac, the most popular of the medicines known to raise serotonin levels.

Because the brain cannot absorb serotonin from the bloodstream, serotonin itself cannot be used. Therefore, researchers have explored the possibility of using serotonin's precursor, 5-HTP, instead of drugs. According to W. Poldinger, M.D., and colleagues at the Psychiatric University Hospital in Basel, Switzerland: "When faced with a deficiency of serotonin in the brain, interfering with the inactivation process is perhaps not quite so obvious a solution as direct substitution of what is lacking. If a substitution therapy [taking serotonin itself] is not feasible, why not take the closest precursor crossing the blood-brain barrier and let the brain itself do the finishing step?" (Poldinger, 1991)

Why not take serotonin instead of 5-HTP?

This is a question I'm often asked by patients or during radio interviews and lectures. If serotonin is the important brain chemical, why not give it directly instead of its precursor 5-HTP? The answer is that serotonin is not able to easily cross the blood-brain barrier. 5-HTP can cross this barrier with great ease.

What are some of the claims being made about 5-HTP?

Do a search on the Internet for 5-HTP, or read health magazine articles or ads promoted by certain vitamin companies, and you'll come across a number of claims that will make you think 5-HTP is the answer to all that ails us human beings, claims such as:

- Your brain will think you've had a Hawaiian vacation.

- What do depression, insomnia, anxiety, suicide, migraines, PMS, obsessive-compulsive behavior, stress, obesity, and addiction have in common? All of these conditions are signs of low serotonin levels. Now that tryptophan has been banned, researchers have clinically investigated 5-HTP in comparison with antidepressant drugs. The results of the studies are astounding.

- If you are unhappy with the results or side effects of your current diet program, 5-HTP is the answer.

- 5-HTP is being used for Parkinson's disease, myoclonus, depression, Alzheimer's disease, anxiety disorders, autism, dementia, and more.

- 5-HTP can help you overcome addiction and pain.

- It's the natural alternative to Prozac.

This book will examine some of these claims.

Do "experts" know everything?

As 5-HTP becomes more popular, you will come across various opinions in the media from people regarded as experts on both tryptophan and 5-HTP. However, are the opinions of all experts reliable? For instance, when I was writing about melatonin, I mentioned to a pineal gland expert my clinical finding that melatonin causes vivid dreaming. He was skeptical and told me that there was no evidence in the medical literature that melatonin causes vivid dreams. Anyone who has taken melatonin and has experienced the intense dreams doesn't need a double-blind study to be convinced that this pineal hormone makes dreams especially lucid.

Then, when I started my book on DHEA, I interviewed the world's top twenty experts in the field. Almost all of them gave me different viewpoints as to the safety and practical uses of this hormone. Their viewpoints often differed diametrically. Furthermore, none of these experts told me that high-dose DHEA use could lead to accelerated hair loss or, even worse, palpitations and heart irregularities. I discovered these effects from my interviews with users of DHEA.

Again, when I wrote about pregnenolone, the "mother of DHEA," I asked one of the leading experts in the field about the ideal time to take this hormone. He told me that I should take it at night before bed. I did, and stayed awake most of the night. Needless to say, I was quite irritable the next morning. After a few days of self-experimentation, I discovered that morning, or before noon, was the best time to take pregnenolone.

These experiences reinforced my assertion that in order to learn about a supplement, I had to find as much information as I could gather from various sources. I also had to experiment on myself.

CHAPTER 2

THE AUTHOR'S 5-HTP DIARY

5-HTP was introduced to the over-the-counter market in 1994. After coming across all sorts of claims about its effects for more than two years, I finally decided to find out for myself whether any of them were true. You may be surprised at what I've discovered.

If I am going to recommend a supplement to my patients and the public, it is only fair that I first try it myself to see whether it is safe and effective. But there's another reason: I have consistently discovered responses to various supplements by taking them myself that had not been previously mentioned in the medical literature. For example, there was the vivid dreaming produced by melatonin, a response that my patients also experienced, and the enhanced visual and auditory perception, along with an enhancement of awareness, produced by pregnenolone. Therefore, I took 5-HTP on numerous occasions over a period of several months.

Works as a sleeping pill

My self-test began in the fall of 1997, while I was attending a medical conference where I had given a talk. One of the vitamin companies promoting 5-HTP was exhibiting at the conference, and I was able to obtain a bottle. I took a 50-mg pill at about 10 P.M., and within ten minutes noticed a yawn. However, I was sitting in a comfortable chair in the hotel lobby, and was able to keep a conversation going with a friend for an hour or so before excusing myself. During this interval, I must have yawned at least every five minutes. I slept very well that night. This very first experience indicated that 5-HTP indeed has sleep-inducing qualities.

During the following weeks, I started coming across more and more positive claims about 5-HTP through the Internet and articles in health magazines. So, I decided to see for myself if any of these claims were true. I went to the University of California Biomedical Library and did a complete literature search for all articles published on 5-HTP in the last three decades. I came across some that did show 5-HTP to have antidepressant qualities, and also learned that no fatalities had ever been reported, even when very high dosages of up to 9,000 mg a day were used (Wyatt, 1973). I decided to do more of my own research.

Daytime sedation

I took 100 mg of 5-HTP at noon on a Sunday, expecting it to have a mood-elevating effect. But, surprisingly, all I wanted to do an hour later was to go to bed and sleep, which I did for an hour. As I lay in bed, I started wondering how this nutrient was supposed to act as an antidepressant if all it did was induce sedation.

To find out, I repeated the same experiment the next day. My expectation was that the 5-HTP would again make me relaxed and sleepy. However, within an hour or so of taking it, I felt alert and focused. This puzzled and frustrated me. It didn't make sense . . . until I recalled that an hour before taking the 5-HTP, I had met a few friends at a local cafe and had two cups of strong coffee, a beverage I normally do not drink.

At about 1 P.M., my normal lunchtime, I realized that my appetite just wasn't there. I also noticed a mild nausea, which came and went. Nausea when doses of 100 mg or more are taken has been frequently reported in the medical literature. In fact, after sedation, nausea is probably the most common side effect from the use of high-dose 5-HTP. I wondered if an antinausea herb could be taken with the supplement: Ginger came to mind. I had by now read a study that 5-HTP had been tested as a weight-loss nutrient (Ceci, 1989), and speculated that its mechanism of action was most likely as an appetite-suppressing agent. In the study, the patients had been given 300 mg of 5-HTP half an hour before eating, three times a day. This dosage seemed excessively high to me, and I would think many of the subjects in the trial would have felt sleepy all day. The article, though, did not mention anything about the subjects feeling sedated.

The next day, I again took 100 mg of 5-HTP at about 4 P.M., this time *without* drinking coffee. An hour later, I felt sleepy, but took another pill to see what would happen. At 6 P.M., I had a very strong urge to sleep and napped for an hour. I awoke feeling refreshed and stayed alert until I went to bed at midnight.

The idea that daytime sleepiness was a shortcoming of 5-HTP use was reinforced the next day, when a patient called to say that the 50 mg she had been taking three times a day was making her feel tired and sleepy all day long.

At this point, I was starting to suspect that in order for 5-HTP to be an effective daytime mood-elevating agent, it had to be combined with a supplement or medicine that counteracted the sedation. Otherwise, few people would continue taking 5-HTP during the day. This seemed especially true when I recalled my experience with caffeine and the appetite suppression I had noticed at that time. Perhaps a low dosage of 5-HTP could be combined with a low dosage of St. John's wort, an herb that has antidepressant properties. I thought that green tea, guarana, tyrosine, ginkgo, and other natural alertness-causing agents, used in low dosages, could be other combination options.

Mild mood elevation and relaxation

After my next nighttime dose of 5-HTP, I felt very alert when I awoke. As I was working in my office, I noticed an elevation of mood that lasted most of the morning. I took 20 mg of 5-HTP at midday and another 20 mg in the late afternoon. I felt slightly sedated without any mood elevation, but did not have a strong urge to sleep. My evening dose was 20 mg at 11 P.M., and I went to bed at midnight and slept soundly until 7 A.M.

I awoke very refreshed. My morning dose was again 20 mg. Later in the morning, I felt extremely relaxed and, even though there were a lot of stressful events going on in the office, I felt little anxiety. The relaxation I had experienced on the low daytime dose made me think that perhaps 5-HTP had a role to play in anxiety disorders. Could it also help children with hyperactivity, which is also called attention deficit hyperactivity disorder (ADHD), become a little calmer? At 11 P.M., I took a 100 mg pill and retired to bed an hour later.

The nightmare

I awoke at 7 A.M. recalling an incredibly vivid dream. It was one of the scariest dreams I had experienced for quite a few months. In the dream, my office assistant and I were chased by gun-wielding criminals through the high desert of eastern California. Bullets were flying in all directions. Our car swerved down mountain roads until we ran out of gas, and we had to flee on foot through sagebrush and cacti. We finally found an isolated house in the remote desert and knocked on the door. We told the woman who lived there about our flight from the vicious criminals, and she said we seemed paranoid and reassured us there was nobody pursuing us. It felt like I had experienced a brief paranoid psychotic episode in my dream, and that perhaps we weren't being pursued after all.

Surprisingly, despite the nightmare, I was in a good mood that morning. For some strange reason, I could shrug off the terrible dream. I did notice, though, that my jaw muscles were sore, perhaps from grinding my teeth in my sleep.

Well, this nightmare made me realize that it was not a good idea to take 5-HTP at a dose of 100 mg. It reminded me of a very unpleasant vivid dream I once had on 5 mg of melatonin, which is a fairly high dose of that supplement. I imagined consumers of 5-HTP were also likely to have these vivid dreams, as many have had when taking melatonin. When melatonin first came on the market, 3- and 5-mg dosages were the most commonly found in stores. Later, doctors learned that low dosages, from 0.3 to 1 mg, were also effective, perhaps even more than the higher ones. I now thought that the 100 mg of 5-HTP being sold was probably too high a dosage, and that many people were going to have the side effects that I had experienced, including nausea, excessive sedation, and nightmares. (Another side effect I noticed while taking high-dose 5-HTP is reduced sex drive, which reverses within a couple of days after stopping the supplement.) I thought that providing 10- and 25-mg pills for daytime use, and 50-mg pills for nighttime use, would be a cautious approach until we learned more about this interesting nutrient.

The concept of combining 5-HTP
with herbs and supplements

It was one of those stormy *El Niño* days in southern California. The rain was coming down hard, but I just love to walk in the rain. We rarely have physical challenges in our daily lives.

I took 50 mg of 5-HTP at 8 A.M. and walked two miles from Marina Del Rey to a cafe on the Venice beachfront. Hardly anyone was outside except for a few dogs walking their owners. My appetite was good and I ordered eggs, lox, and a bagel with a side of tomatoes. My appetite didn't seem too affected as I finished the whole plate and went back outside in the rain. All the streets in Venice were quiet, and the foliage in the gardens seemed refreshingly lush, so green and alive from the almost nonstop, tropical-like drenching California had received over the previous few weeks. I noticed a slight enhancement of visual clarity.

I came back home a few hours later and had the desire to take a nap, which was unusual for me despite the long walk and big breakfast. I felt very relaxed and sleepy most of the afternoon, almost as if I had been on Valium, with little motivation but to just lie on the sofa by the fireplace. I watched through the balcony window as the storm swayed the trees while the whitecaps crashed on the marina jetty.

While sitting by the fireplace, I had a chance to reflect on my experiences thus far with 5-HTP. I was starting to get a good understanding of its effects. It was definitely sedating, whether taken during the day or at night, although the sleep-inducing aspect was less pronounced during the day. But could it be used as an antidepressant? In order for a supplement or medicine to work well as an antidepressant, it should not cause excessive sedation during the day. Furthermore, if 5-HTP were to be used as an appetite suppressant before meals, the sedation would be a significant limiting factor. I was becoming more and more convinced that the daytime dose of 5-HTP needed to be small so as not to induce sedation, and that combining it with an alertness-causing, or appetite-suppressing, supplement was necessary. However, individuals who were very anxious and tense could probably tolerate daytime 5-HTP very well.

I recalled that the 100 mg I had taken the night before had given me a terrible nightmare. If 5-HTP was going to be used as a sleep aid, lower dosages would need to be used. Perhaps 5-HTP would be a better treatment for insomnia when combined with sedative herbs, such as valerian and hops. These herbs are not known to induce vivid dreaming.

5-HTP really is a mood elevator!

I then decided to take 5-HTP only at night for a few evenings just to see if this would have any influence on me during the day. For one week in a row, I took doses ranging from 25 to 75 mg of 5-HTP about one half-hour to an hour before bed. My sleep was consistently good. I had more dreams than usual, but they were not as vivid as the one I had had on the 100 mg. No further nightmares occurred.

After about three or four days on this regimen, I noticed that my mood was slightly elevated during the day, and I was feeling more calm and focused. I had slightly more energy in the evening, but came 10 P.M., I was starting to yawn, at least an hour or two before my normal bedtime. Had my sleep cycle been shifted back an hour? Another thing I noticed was a slight enhancement of visual clarity, although it was not as pronounced as that caused by pregnenolone.

For a couple of days, I took a multivitamin pill in the morning that included about twice the Recommended Daily Allowance of the B vitamins. I found that the additional vitamins gave me more energy.

At this point, I started thinking that perhaps 5-HTP did have subtle antidepressant qualities after all, although the response to 5-HTP in someone who is clinically depressed may be different than mine. There was certainly a lot more to learn about how to use this nutrient appropriately.

Shedding wanted pounds

After about a week of just taking the 5-HTP at night, and noticing the slight mood-elevating effect during the day, I decided to use a small amount of this supplement throughout the day. I took 10 mg in the morning and early afternoon on an empty stomach. This small amount did not make me feel sleepy, but actually tended to heighten the mood-elevating effect I was already feeling. There was a noticeable decrease in appetite. Eating did not seem as high a priority, especially when I really got involved in my work. I would sit for hours in front of the computer and forget to eat. I lost three pounds in a short period of time, even though I was trying to maintain my weight.

Tolerance to 5-HTP

Then it happened: I had continued taking the 5-HTP every evening and was starting to wake in the middle of the night without getting back to sleep. This had happened to me in the past when I had taken melatonin regularly. Perhaps my brain was adjusting to the regular use of supplemental 5-HTP. I stopped taking it for a few days, after which I began to use it twice a week. I still continue to do so, and have not developed a tolerance. The tolerance to nightly use prompted me to develop an insomnia treatment plan that uses alternating sleep-inducing agents, as discussed in Chapters 7 and 11.

I now tried taking 5-HTP during the day and noticed that it wasn't making me sedated as it had in the past. In fact, I would often feel more alert and focused. After a few more days of experimentation, it seemed that I did not feel sleepy on the days when I was already very alert. However, on days when I felt a little sluggish, due perhaps to not having slept too well the night before, taking 5-HTP during the day induced the urge to sleep.

Since those early days of experimentation, I have taken 5-HTP on numerous occasions and in various dosages, either by itself or combined with other medicines, herbs, nutrients, and hormones. You can read more about my personal experiences throughout the book, and about the experiences of others, including doctors who use 5-HTP in their practices.

Before seeing how 5-HTP can be used in the treatment of specific disorders, let's first explore the connections among tryptophan, 5-HTP, and serotonin.

CHAPTER 3

THE TRYPTOPHAN TO MELATONIN CHEMISTRY TRIP

T here are dozens of chemicals that coordinate the normal functioning of our brains. Some of the best-known brain chemicals—which are also called *neurotransmitters*— are serotonin, dopamine, norepinephrine, acetylcholine, and gamma-amino-butyric acid (GABA). Serotonin happens to be the most studied neurotransmitter, since it helps regulate a wide range of psychological and physical functions.

Although serotonin may be obtained from a variety of dietary sources, the amounts generally ingested are not adequate, and almost all of the serotonin the body needs is made from the amino acid tryptophan. As we have seen, the body turns tryptophan into 5-HTP, which in turn is used to create serotonin. In this chapter, I will first discuss what serotonin is, and how it, and other neurotransmitters, work. I'll then explain 5-HTP's role in this process. (For more detailed information, see "A Note to Professionals" in Appendix A.)

What is serotonin?

Serotonin—also known as 5-hydroxytryptamine, or 5-HT—is widely distributed in animals and plants, especially in certain fruits and nuts. It was first isolated from blood in 1948 (Rapport, 1948). It was later found in the intestinal wall, where it causes increased gastrointestinal motion, and blood vessels, where it plays a role in constricting large vessels. Platelets, those small cells in the blood that help form clots, also contain serotonin.

Perhaps serotonin's best-known function is that of a brain chemical, despite the fact that only from 1 to 2 percent of the body's entire supply is found in the brain. The wide extent of psychological functions regulated by serotonin involves mood, anxiety, arousal, attention, aggression, and thought. Disruption of serotonin's normal functioning leads to a number of psychiatric conditions, including anxiety disorders, depression, alcoholism, violence, improper social behavior, and sexual aberrations. Common medical conditions associated with such disruption include chronic pain, obesity or eating disorders, and disturbances in the sleep-wake cycle. These effects result from serotonin's actions as a neurotransmitter.

Neurotransmitters: messengers of the brain

Brain cells, also called *neurons*, are not in direct contact with one another, but are separated by spaces called *synaptic clefts*. Neurotransmitters, such as serotonin, pass messages from one neuron to another across these clefts. Serotonin is made in neurons and stored in small enclosures called *vesicles*. When the neuron receives a nerve impulse, the vesicles open, and serotonin is released into the synaptic cleft. At that point, the serotonin is picked up by serotonin *receptors* located on adjoining neurons, and the message is passed (see the diagram on the next page). Scientists have discovered that there are at least seven different types of serotonin receptors, and that these types have their own subtypes.

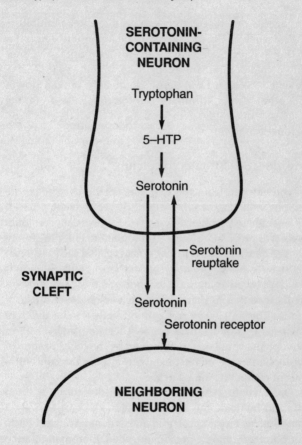

While the serotonin is in the synaptic cleft, it can be either taken back into the neuron that released it in the first place, a process called *reuptake*, or broken down by enzymes located within the synaptic cleft. In general, the major mechanism that ends serotonin action in the synaptic cleft is its reuptake.

Various drugs can affect serotonin levels. Antidepres-

sant agents called selective serotonin reuptake inhibitors
(SSRIs), such as Prozac, increase the amount of serotonin
in the synaptic cleft by preventing reuptake. Other
drugs, called monoamine oxidase inhibitors (MAOIs),
prevent serotonin from being chemically broken down.
Still other drugs stimulate serotonin release from the
vesicles, while others either mimic or block its effects on
various receptors.

How does 5-HTP enter the picture?

As you can see from the diagram on the next page, the
building material, or *precursor*, of 5-HTP is tryptophan. If
not enough tryptophan is supplied to the brain, serotonin
levels will drop. In order for tryptophan to enter the brain,
it has to be transported across what is called the *blood-brain
barrier* by a carrier protein. This carrier protein, though, is
also used by other amino acids. Imagine this carrier to be a
small canoe that can carry only one person across a lake at
a time. There's always competition between the different
amino acids to jump on the canoe. Therefore, brain levels of
tryptophan are not only determined by the concentration of
tryptophan in the bloodstream, but also by the concentrations of competing amino acids.

Once tryptophan enters the brain, it can go to brain
cells and be made into 5-HTP. There is an enzyme called
tryptophan hydroxylase (TH) that makes this conversion. From
this point on, another enzyme, called L-aromatic amino
acid decarboxylase, or *AADC*, takes over to change 5-HTP
into serotonin.

Tryptophan hydroxylase is present in low concentrations in most tissues, including the brain (Cooper, 1996).
AADC is found in many parts of the body, including the
nervous system, liver, digestive system, and other tissues,
even the kidneys (Wa, 1995). Because there is a greater

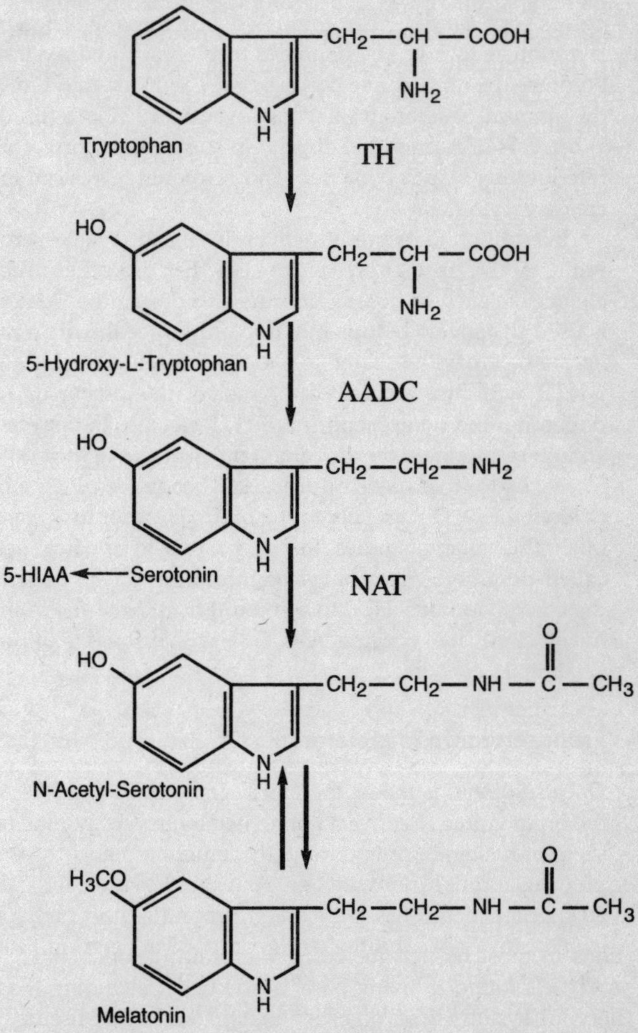

CH2—CH—COOH
NH2
Tryptophan
TH
HO
CH2—CH—COOH
NH2
5-Hydroxy-L-Tryptophan
AADC
HO
CH2—CH2—NH2
5-HIAA ← Serotonin
NAT
HO
O
CH2—CH2—NH—C—CH3
N-Acetyl-Serotonin
H3CO
O
CH2—CH2—NH—C—CH3
Melatonin
N
H

abundance of AADC, the conversion of 5-HTP to serotonin occurs very rapidly when compared with the conversion of tryptophan to 5-HTP. Therefore, tryptophan hydroxylase becomes the *rate-limiting enzyme*, or the enzyme that limits the amount of serotonin that can eventually be produced. When 5-HTP is provided directly in supplement form, this rate-limiting step is bypassed, and serotonin levels can increase very quickly.

In addition to its presence in brain cells that make serotonin, AADC is also present in cells that use other brain chemicals, such as norepinephrine and dopamine. AADC is used to convert L-dopa into dopamine (see the diagram on the next page). In later chapters, I'll discuss combining 5-HTP with the amino acid tyrosine, the precursor of dopamine and norepinephrine. 5-HTP can also be converted into serotonin in the blood and other places in the body. However, for therapeutic purposes, it would be preferable to have the 5-HTP go only to the brain. In order to accomplish this, many studies in the past used medications called peripheral decarboxylase inhibitors (PDIs) to stop the conversion of 5-HTP to serotonin outside of the brain by blocking the enzyme AADC. I will discuss PDIs in Chapter 11.

From serotonin to melatonin

Once serotonin is made, the pineal gland in the middle of the brain is able to convert it into melatonin. Melatonin is a sleep-inducing hormone, and the amount made by the pineal gland declines with age. Some studies have indicated that melatonin has antioxidant properties and perhaps even antitumor abilities. (For more information, see *Melatonin: Nature's Sleeping Pill* in Appendix B.)

As you can see from the diagram on page 27, serotonin goes through an intermediary step before turning into mela-

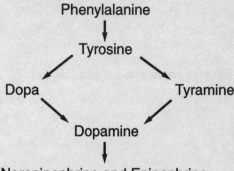

Phenylalanine

Tyrosine

Dopa   Tyramine

Dopamine

Norepinephrine and Epinephrine

tonin, becoming N-acetyl-serotonin, or NAS. This step is assisted by a rate-limiting enzyme called N-acetyl-trans-ferase, or *NAT*. During daylight hours, or during exposure to any kind of bright light, the activity of NAT slows dramatically as the result of signals relayed from the eyes to the pineal gland. At night, or during periods of darkness, NAT is much more active. Therefore, according to Bob Perry, Ph.D., who has done research on how serotonin becomes melatonin, the pineal gland will not be able to make melatonin during the day after 5-HTP is consumed because the NAT enzyme has been mostly, or completely, inactivated by light.

However, it is known that the ingestion of high doses of 5-HTP can lead to daytime sedation and the urge to sleep. During my discussion with Dr. Perry, we thought this could be due to sedation caused by serotonin alone, or to that caused by serotonin being converted into melatonin in the digestive system, since there are specialized cells in the intestines that can make melatonin. Melatonin that is made in the intestines normally stays within its local environment and generally does not enter the regular blood circulation. Perhaps by ingesting a high dose of 5-HTP during the day, enough melatonin can be made in the intestines that it can spill over into the general circulation, enter the brain, and induce sleep. (For more information on 5-HTP and sleep, see Chapter 7.)

Why not take tryptophan instead of 5-HTP?

Tryptophan was available over the counter until 1989, when it was taken off the market because of a contamination problem (see Chapter 1). Tryptophan is now available only through compounding pharmacies and only by prescription. The tryptophan ban does not apply to 5-HTP, which lessens the impact of the ban.

In addition to problems with availability, there are important differences between tryptophan and 5-HTP. As I explained earlier, 5-HTP is more readily converted into serotonin. These substances also differ in their rates of absorption and in the ways they are processed within the body. (For a more detailed explanation, see Appendix A.)

Anecdotal information indicates the amount of 5-HTP needed to induce sleep is generally a fifth to a tenth of a required dose of tryptophan. For instance, if it normally takes about 500 mg of tryptophan to induce and maintain sleep, only about 50 to 100 mg of 5-HTP is needed. This conversion ratio of 5 to 1 or 10 to 1, tryptophan to 5-HTP, is a rough guideline, and each individual would have his or her own ratio. This could even fluctuate from day to day, depending on food and vitamin intake, concurrent use of medicines, age, and, in women, menstrual cycle.

The actions of tryptophan are in many ways similar to those of 5-HTP, although they have rarely been tested head to head. I have interviewed users who prefer tryptophan, while others prefer 5-HTP. You could certainly try each nutrient separately to see which best suits your needs.

The carbohydrate-tryptophan-serotonin connection

Brain serotonin levels depend both on the amount of tryptophan taken in through the diet and on blood levels of tryptophan relative to those of the other amino acids. When volunteers drink an amino-acid mixture containing all the essential amino acids except tryptophan, there's a rapid decrease in brain serotonin levels (Smith, 1997). This is called the tryptophan depletion test. People who already have a tendency towards depression and anxiety become more depressed and anxious (Young, 1996).

If you wish to temporarily raise brain levels of tryptophan, and hence those of serotonin, would it be better to eat a protein meal that contains tryptophan, or a carbohydrate meal, such as pasta, that does not have much tryptophan? The answer, surprisingly, is the latter. You are probably thinking that it doesn't make sense. There is a good explanation, though.

Foods that contain protein, such as turkey, fish, or other types of meat, contain tryptophan, but they also contain an abundance of the other twenty-two amino acids. Once ingested, a balance will exist between the amount of tryptophan in the blood and amounts of the other amino acids. When a meal is ingested consisting mostly of carbohydrate, the elevation in blood-sugar levels stimulates the release of insulin. Insulin encourages amino acids such as valine, isoleucine, and leucine to leave the blood, and enter many tissues and organs (Wurtman, 1996). Hence, there is a higher level of tryptophan in the blood compared with levels of the other amino acids. This tryptophan can now be shuttled more easily from the blood to the brain, where it will eventually be converted into serotonin.

The serotonin deficiency that occurs in premenstrual syndrome and depression may be the reason why carbohydrate craving is common in these conditions. Over the long run, of course, it is better to eat meals that are balanced for proteins, fats, and carbohydrates, since raising blood-sugar levels consistently is not healthy.

Having learned the basics about 5-HTP and serotonin, we can now examine some of the research done with 5-HTP, and how it can be used clinically in a variety of medical and psychiatric conditions.

CHAPTER 4

APPETITE CONTROL— EAT LESS, WEIGH LESS

Brain levels of certain neurotransmitters and hormones influence appetite. Can we safely manipulate these neurotransmitters and hormones in order to induce satiety, and thus reduce caloric intake? There are tens of millions of overweight Americans who would love to know the answer to this question. But since the Phen-Fen and Redux scare, when FDA-approved diet drugs were found to have serious side effects, the public and the media have been much more cautious about jumping on the bandwagon when a new diet pill is introduced. This caution was apparent when, a year after Redux entered the market, it was announced that a new diet drug called Meridia would soon be introduced. Members of the media were much less enthusiastic about promoting Meridia.

But is there a safer alternative to these diet drugs? Is 5-HTP the new wonder, natural weight-loss pill? Perhaps. Patients to whom I have recommended 5-HTP for appetite control have been very satisfied with the results thus far. Nancy, a 52-year-old patient, tells me, "My weight is dropping off amazingly. I haven't been this thin in ten years. I normally weigh between 138 and 142 pounds. I started taking 5-HTP six weeks ago at a dose of 25 mg twice a day. I'm now down to 130 pounds. I also feel calmer and I'm able to work better since I'm not as anxious."

One of the most important neurotransmitters involved in satiety is serotonin. Pharmacological, biochemical, and behavioral evidence accumulated over the past three decades suggests that serotonin tends to inhibit eating behavior (Li, 1983; Lieberman, 1986). Short-term observations have shown that 5-HTP, even without dietary restrictions, causes loss of interest in food, decreased food intake, and weight loss in obese individuals. I have personally noticed appetite suppression as a result of 5-HTP use, and so have many of my patients.

In this chapter, we will first look at research done on 5-HTP and weight loss. I will then discuss the pros and cons of 5-HTP before telling you about other natural weight-loss supplements that can be used with this important nutrient.

Research on 5-HTP and weight loss

Several studies have been done to evaluate the role of 5-HTP in weight control.

The 1989 Italian study

Researchers at the Department of Internal Medicine, University of Rome, La Sapienza looked at nineteen obese women in a study aimed at evaluating the effects of 5-HTP on feeding behavior, mood state, and weight loss (Ceci, 1989). It was a double-blind crossover study, which simply means that neither the researchers nor the volunteers knew who was taking the nutrient and who was taking a placebo, and that the women switched substances in the course of the study. The women, whose average age was forty, tended to consume more calories a day than they required.

During two five-week periods, either oral 5-HTP—at a rate of 8 mg per kilogram of body weight per day—or a placebo was administered. For a 60-kg woman (about 132 pounds), the average daily dosage was about 480 mg a day. This was given in three divided doses, one taken about thirty minutes before each meal.

No dietary restrictions were applied, and the patients were seen weekly to evaluate their feeding behavior, mood, and body weight. Food intake was determined using a three-day diet diary. In order to make sure the patients were taking their medicine, urine tests were done regularly to check for levels of 5-hydroxy-3-indole acetic acid (5-HIAA), which is produced when serotonin is broken down by the body.

The use of 5-HTP resulted in no reported changes in mood but did lead to decreased food intake. Fifteen of the nineteen women reported appetite loss during 5-HTP therapy, but only six women reported appetite loss while on the placebo. The average daily caloric intake before treatment was 2,900 calories. While the women were on the placebo pill, their intake dropped to 2,300 calories, indicating that just being monitored on a weight-loss program can motivate people to eat less. However, caloric intake dropped to 1,800 when the women took 5-HTP. The average weight loss during the five-week period for the 5-HTP group was about 2.2 kg, or close to 5 pounds. Common side effects reported by the patients on 5-HTP were nausea and vomiting, which occurred in 60 percent of the women.

The researchers conclude, "The results of this study are in favor of a specific role played by the serotonergic system [serotonin] in the regulation of feeding behavior in humans."

One can presume from this preliminary study that gastrointestinal side effects would limit the use of high-dose 5-HTP in appetite control. Giving a lower dose of 5-HTP would lessen the rate of side effects. The resulting weight loss may be slower, but in the long run the nutrient would be better tolerated.

The 1992 Italian study

This study was also done at the Department of Internal Medicine, University of Rome, La Sapienza, and had a similar protocol, or study method, as the 1989 study, except that it went on for two consecutive six-week periods. Twenty obese patients were randomly assigned to receive either 300 mg of 5-HTP or a placebo three times a day, thirty minutes before each meal (Cangiano, 1992). No particular diet was prescribed during the first period, and the second period required the subjects to moderately reduce their caloric intake. There was significant weight loss in the patients who received 5-HTP during both periods: They ate less carbohydrate and reported feeling more satiated. A commonly reported side effect was nausea.

The researchers state,

> *The results of this study seem to confirm the specific role played by the serotonergic system in the regulation of feeding behavior in humans. The administration of 5-HTP was in fact followed by a significant loss of body weight. The optimal adherence to dietary prescription as well as the good tolerance to 5-HTP treatment observed suggest that this substance may be safely used in the long-term treatment of obesity.*

The pros and cons of 5-HTP

It's always been a challenge for doctors to treat obesity. Over the past few years, many have relied on the quick fix by prescribing diet drugs. A number of these drugs influence serotonin, and thus help promote weight loss through appetite suppression. We now have a natural serotonin precursor that could, theoretically, achieve the same goal. However, 5-HTP is not perfect. Let's discuss some of its advantages and disadvantages.

5-HTP has the following advantages:

- It will naturally raise levels of brain serotonin, the "satiety neurotransmitter."

- The dosage can be adjusted up or down as needed.

- It can easily be taken on an empty stomach.

- It appears, at least in the short run, that there are no significant side effects in low dosages.

- It is not currently known to cause high blood pressure, heart irregularities, or nerve cell damage, as has been the case with some diet drugs.

5-HTP has the following disadvantages:

- When taken during the day, high dosages of 5-HTP can, in some individuals, lead to sedation. In order to prevent this sedation, it should either be taken in small dosages, or be combined with an alertness-causing supplement.

- 5-HTP is not as expensive as diet drugs, but it still can cost the consumer a good penny.

- Nausea is a common side effect on higher dosages.

- Although 5-HTP is safe for short-term consumption, no thoroughly evaluated long-term studies, meaning those that have continued for more than one year, have been done to determine its potential risks.

• It's possible that tolerance could develop. While 5-HTP may work in the short run, we don't know if the benefits will continue in the long run.

Considering these factors, it would seem prudent to take the least amount of 5-HTP that is effective, and to take breaks while using it. In addition, this nutrient should be combined with other weight-loss, or appetite-controlling, supplements. See Chapter 11 for a step-by-step program.

Additional weight-control supplements

Even though 5-HTP appears promising as an antiobesity nutrient, it is quite likely that we would get better results if we used it with other natural supplements that have different, or complimentary, modes of action.

There are a number of products on the market that are promoted as weight-loss agents. These include chitosan, which is purported to bind fat in the intestines so the fat can be excreted before absorption; the herb *Garcinia cambogia* and the nutrient pyruvate, touted as fat burners; and the fatty acid CLA (conjugated linoleic acid). At this point, there aren't enough studies with humans to convince me that any of the above substances is truly effective and worth the expense. It's possible, though, that future research may show that one or more of these supplements does work.

There are some supplements that have shown promise as parts of a balanced weight-loss program. These include fiber and natural stimulants.

Fiber

Psyllium powder, also known as plantago ovata husk, is a

natural fiber available over the counter in health food stores, retail outlets, and grocery stores. It is sold under the generic name "psyllium" or under a variety of brand names, and is often flavored with orange or lemon. Its most common use is as a bulk laxative to relieve constipation, since it has the ability to absorb water and make stools softer. Studies have also shown that psyllium has the ability to bind to cholesterol and other fats, and carry these substances out of the body. Thus, the regular use of psyllium, as part of an overall cholesterol-lowering regimen, helps lower lipid levels in the blood.

What is less known about psyllium is that it can be helpful in reducing weight by decreasing hunger. Researchers at the University of London performed a double-blind crossover study evaluating the role of psyllium supplements on appetite (Turnbull, 1995). Seventeen young women were involved in the study. Each was given either psyllium in a glass of water before lunch, or a similar-looking placebo, or plain water. After lunch, the women were asked to answer a questionnaire about the meal. There was a significant difference between the psyllium group and the placebo group. The women who had ingested the psyllium reported feeling fuller, and their total fat intake was lower. The reason is most likely due to psyllium's absorption of fluid in the intestines, which would promote a feeling of satiety.

The FDA has since announced that manufacturers of cereals and other foods containing psyllium can claim that their products reduce the risk of heart disease when combined with a diet low in saturated fat and cholesterol.

Other fibers can work well, too, including pectin and guar gum. Pectin is a complex carbohydrate that cannot be absorbed from the intestinal system. It is used in the food industry as a thickening agent, and in the production of jams and jellies. Pectin has the ability to induce satiety in obese individuals (DiLonrenzo, 1988).

To test whether this safe effect could be seen in healthy subjects, researchers in the United States Army at Fort Sam Houston, Texas, gave pectin to seventy-four Army volunteers, forty-nine men and twenty-five women. On two separate days, these volunteers fasted overnight and then drank 450 ml of orange juice followed four hours later by 470 ml of ice cream. On one of the two days, 5, 10, or 20 grams of pectin were mixed with the orange juice. Satiety was measured before ingestion; at zero, one, two, three, and four hours after the orange juice; and at zero, thirty, and sixty minutes after the ice cream. The results showed that pectin was effective in all dosages. The effect lasted up to four hours after ingesting pectin and orange juice, and for sixty minutes after a second meal consisting of ice cream. The researchers say, "Pectin in doses as small as five grams mixed with orange juice increases satiety and can aid in a program to reduce weight by limiting food intake."

A one-week supplementation of 40 grams a day of guar gum in obese women was also found to reduce caloric intake (Pasman, 1997). As you can see, there are a number of fiber types that can play a role in obesity treatment.

Natural stimulants

Caffeine in coffee, tea, and guarana may work together with 5-HTP to induce appetite suppression, although one should avoid excessive consumption of caffeine-containing products. Other stimulants include St. John's wort, and the amino acids phenylalanine and tyrosine.

Some vitamin companies have marketed a controversial herbal appetite suppressant called ma huang, the Chinese name for the herb ephedra. Promoted for its energy-enhancing and weight-loss possibilities, ma huang can be found in health food stores under several brand names. It is also marketed in combination with other ingredients, including caffeine, green tea extract, kola nut, guarana, and St. John's wort. These products are promoted as "energy pills" or "diet pills."

Ma huang has been used in China for over 5,000 years for a variety of conditions, including asthma and hay fever. It was not until 1923 that scientists discovered the active ingredients found within this herb: ephedrine and pseudoephedrine. Both of these chemicals are central nervous system stimulants. Ephedrine constricts blood vessels and stimulates the heart. Pseudoephedrine enhances the dilation of bronchial tubes in the lungs and helps relieve nasal congestion. There also is some evidence that compounds found in ma huang can play a role in weight loss when used as part of a calorie-restricted diet (Toubro, 1993).

However, some people may not respond well to ma huang. Anyone with cardiovascular problems should not take this herb, since excessive consumption of ma huang can lead to high blood pressure, serious heart irregularities, and even death. At this time, I don't recommend that you use ma huang unless you are closely supervised by a health care provider.

Summary

In recognizing the important biochemical role of 5-HTP as a serotonin precursor, it appears that 5-HTP supplements can be helpful in controlling appetite, at least for the short term. This nutrient could potentially prove to be a safer alternative to some diet drugs. Until we learn more about the safety of 5-HTP, though, it should only be used temporarily, although taking it in cycles may be an option.

CHAPTER 5

LISTENING TO 5-HTP–
THE SEROTONIN SOLUTION
TO DEPRESSION?

Mood depends on the levels and interactions of a number of different chemicals within the brain, including serotonin, norepinephrine, and dopamine (Larisch, 1997). Depression is mostly due to an imbalance in or shortage of some of these chemicals, which are also known as neurotransmitters.

In this chapter, I will first explain the various types of depression. I will then discuss the various types of chemical antidepressants, since these drugs are in such wide use. Finally, I will present a multisupplement therapy that can help you fight depression naturally, along with some thoughts on possibly combining 5-HTP and prescription antidepressants.

The lowdown on depression

There are many types of mood disorders, and their causes are numerous. Many people have a genetic predisposition that leads to deficiencies in certain brain chemicals or hormones—you may know of certain families in which depression is rampant. Depression can also occur due to a particularly stressful life event, such as divorce, accident, or death in the family. This is called *reactive depression*, and can often last for months or years. Common types of depressive disorders include major depression, dysthymia, manic-depressive illness (also known as bipolar disorder), and seasonal affective disorder (SAD). In addition, depression is a component of premenstrual syndrome, which is covered in Chapter 9.

Major depression—This is a serious disorder that can disrupt a person's entire life, interfering with work and relationships. Generally, symptoms include sadness, hopelessness, and low self-esteem. However, different people show depression in different ways, and some depressed individuals become agitated and anxious. Other symptoms include insomnia, problems in thinking clearly, loss of sexual desire, and fatigue. Depression can even be fatal if it leads to suicidal thoughts. Extremely low levels of serotonin are associated with impulsive behavior and suicide attempts (Linnoila, 1992).

Dysthymia—This is a mild form of depression that is extremely common. Individuals with this condition live normally, hold jobs, and do everything else required by society, but they have little excitement about life and get little joy from day-to-day activities. Low levels of serotonin may be one of the reasons for this low mood. Treatment with Prozac, a drug that affects serotonin levels, has benefited some patients with dysthymia, indicating that serotonin does indeed play a role in this condition (Vanelle, 1997).

Manic-depressive disorder—This illness is marked by depression that alternates with periods of mania. During the manic phase, there is elation with hyperactivity, increased irritability, and little need for sleep. The overenthusiastic quality of the mood and the expansive behavior initially attract others, but the irritability, mood swings, aggressive behavior, and grandiosity in most cases lead to marked interpersonal difficulties. Symptoms start in late teens and early adult life. Lithium is an effective therapy for manic-depressive disorder, as is divalproex (Bowden, 1996). This disorder may have a genetic basis, and recent studies have shown that some patients may have abnormalities in a gene that codes for tryptophan hydroxylase—the enzyme responsible, as we saw in Chapter 3, for converting tryptophan into 5-HTP (Bellivier, 1998).

Seasonal affective disorder (SAD)—This form of depression occurs in wintertime, especially in northern climates where daytime light is of short duration. The depression is due to decreased exposure to full-spectrum sunlight. Symptoms include low mood, poor motivation, and lack of energy. In one study, results indicate that SAD is associated with abnormalities in serotonin metabolism (Levitan, 1998).

Pharmaceutical antidepressants and light therapy are known to be effective against SAD. The problem with light therapy is that it is too time-consuming. Fortunately, it appears that St. John's wort, an antidepressive herb, works just as well. For four weeks, a group of sixty individuals with this disorder was given 300 mg of St. John's wort three times a day (Martinez, 1994). The results were convincing: The herb was effective in lifting mood and relieving symptoms. 5-HTP has not been tested as a treatment for SAD, but it would be worthwhile to try this serotonin precursor.

The medical approach to depression

Antidepressive drugs were first introduced back in the 1950s. Since then, a number of different antidepressants have been created. Many of these medicines have enormously benefited countless patients by relieving despair and suicidal urges. As we learn more about the biochemical basis for depression, we are able to fine-tune our approach and use medicines that provide the most benefits with the fewest risks.

The three most common types of antidepressants are the tricyclics, the monoamine oxidase inhibitors (MAOIs), and the selective serotonin reuptake inhibitors (SSRIs). The tricyclics and MAOIs influence levels of all three major brain chemicals involved with mood: norepinephrine, dopamine, and serotonin. The SSRIs work on serotonin almost exclusively. In addition, there are antidepressants that do not clearly fall into any of these classes.

Tricyclics—These were the first significant antidepressants to be developed. They work by elevating levels of several neurotransmitters, which improves mood. This class includes desipramine (Norpramin), nortriptyline (Pamelor, Aventyl), and amitriptyline (Elavil, Limbitrol, Endep). Because the tricyclics have been in existence for so long, their patents have run out, and thus they are relatively inexpensive. Their biggest drawback is that they can produce a number of side effects, including blood pressure problems, drowsiness, digestive complaints, sexual difficulties, and heartbeat irregularities.

Monoamine oxidase inhibitors (MAOIs)—This was the next group of drugs to be developed. Serotonin, norepinephrine, and dopamine are called monoamines. Once they are released into a synaptic cleft between brain cells, the enzyme monoamine oxidase (MAO) breaks them down. Therefore, a monoamine oxidase inhibitor (MAOI) is a drug that inhibits this enzyme, thus allowing for more of these neurotransmitters to stay in the synaptic cleft and elevate mood. The best-known MAOIs are Nardil (phenelzine) and Parnate (tranylcypromine).

MAOIs were used more frequently in the past, and were prescribed when the tricyclics did not work. MAOIs work very well, but are not commonly used because they can interact with certain foods and drinks to cause dangerous side effects, including a dramatic increase in blood pressure. The substance in foods most responsible for this interaction is tyramine, found in aged cheese, wine, and pickled or smoked meats or fish. Since the introduction of the SSRIs, MAOIs are now prescribed even less frequently than before.

There are two types of MAO inhibitors, A and B. Deprenyl, a medicine currently used for Parkinson's disease, acts as an MAOI type B (for more information, see Appendix A). Deprenyl, also known as selegiline, appears under the brand names Eldepryl and Atapryl.

Selective serotonin reuptake inhibitors (SSRIs)—This class includes the popular antidepressants Prozac (fluoxetine), Zoloft (sertraline), Paxil (paroxetine), and Luvox (fluvoxamine). As the name of the class implies, SSRIs mostly influence levels of serotonin. However, they may also slightly influence levels of other brain chemicals, such as norepinephrine and dopamine.

The SSRIs have become the most popular class of antidepressants because they tend to have fewer side effects than either the tricyclics or the MAOIs. But this does not mean that they are completely free of side effects, which can include headaches, gastrointestinal difficulties, drowsiness, insomnia, and reduced sex drive.

Other antidepressants—Some antidepressants do not clearly fall into any of the three main classes. Anafranil (clomipramine), Effexor (venlafaxine), and Serzone (nefazodone) are called serotonin reuptake inhibitors, as opposed to selective serotonin reuptake inhibitors, because their action on serotonin is not as specific as that of the

SSRIs. Wellbutrin (buproprion) and Desyrel (trazodone) are
other drugs introduced in the past few years.

Studies on 5-HTP and depression

Sometimes, it is difficult to predict which type of medicine a depressed person will respond to, since different
types of brain chemicals will influence mood. One Italian
study has indicated that doctors can potentially predict
an individual's response to an antidepressant by first
measuring the ratio of tryptophan to other amino acids in
the blood. Patients who have low levels of tryptophan,
compared with other amino acids in their bloodstreams,
are more likely to respond to SSRIs, since these antidepressants elevate brain serotonin levels (Lucca, 1994). But
can 5-HTP, the serotonin precursor, effectively substitute
for SSRIs?

Let's evaluate a few of the published studies regarding
the therapeutic influence of 5-HTP on depression.

Prior to 1988: the early studies

The very first significant study that evaluated the role of 5-
HTP as an antidepressant was published in Japan in the early
1970s (Sano, 1974). In this study, 107 patients with depression
and manic-depressive disorder were treated for periods lasting from seven to thirty-five days with between 50 and 300 mg
of 5-HTP. Of the 107 patients, seventy-four were either cured
or showed marked improvement. Later studies in Japan also
found 5-HTP to be effective in treating depression, but high
dosages were used, and many patients reported side effects
such as nausea, vomiting, and diarrhea. The common occur-
rence of side effects made Japanese researchers less enthusiastic in pursuing trials with 5-HTP, and attention focused on
pharmaceutical antidepressants instead (Nakajima, 1996).

An excellent review article on 5-HTP was published in 1987 by William F. Byerley, M.D., and colleagues at the University of Utah. The article summarized the two dozen or so studies that had been done globally on the role of 5-HTP in the treatment of depression. After thoroughly evaluating each study and summarizing the findings, the researchers conclude,

> *A review of the investigations of 5-HTP suggests that this compound has antidepressant properties. Although it may have some efficacy in treatment-refractory patients [those who don't respond to other treatments], 5-HTP should probably be more thoroughly tested in patients likely to respond to antidepressant treatment. Oral administration of 5-HTP is associated with few adverse effects.*

1988: 5-HTP versus an MAOI

Researchers from the Department of Biological Psychiatry, located at Psychiatric Centre Bloemendaal, in The Hague, The Netherlands, compared the effectiveness of 5-HTP with that of the MAOI tranylcypromine in an open study involving patients with major depression. These patients had already been treated unsuccessfully with at least two tricyclic antidepressants (Nolen, 1988).

Treatment with 5-HTP was started in twelve patients at a dose of 10 mg twice a day. This was increased to two daily doses of up to 100 mg if symptoms did not improve. Tranylcypromine was started in fourteen patients at 10 mg twice a day. This was increased to two daily doses of up to 50 mg in cases of insufficient response.

The patients treated with 5-HTP did not improve. In contrast, tranylcypromine was effective in 50 percent of the patients. 5-HTP alone, therefore, was not effective in patients with severe depression who had already failed to respond to previous antidepressant therapy.

1989: weight loss but no change in mood

In Italy, researchers at the Department of Internal Medicine, University of Rome, La Sapienza studied the effect of 5-HTP on mood and food intake. Nineteen obese female subjects were included in a double-blind crossover study aimed at evaluating the effects of oral 5-HTP administration on feeding behavior, mood state, and weight loss (Ceci, 1989). "Double-blind crossover" means that neither the volunteers nor the researchers knew who was taking the nutrient and who was taking a placebo, and that the women switched substances during the study. Oral 5-HTP was given at a rate of 8 mg per kilogram of body weight (or about 17 mg per pound) per day, divided in three daily doses, for a period of five weeks.

The administration of 5-HTP resulted in no changes in mood, but did lead to decreased food intake and weight loss. (For more information about the weight-loss aspects of this study, see Chapter 4.)

1991: 5-HTP versus an SSRI

This Swiss study by Dr. W. Poldinger and colleagues was designed as a double-blind, multicenter trial in patients randomly assigned to receive either Luvox (fluvoxamine), an SSRI, or 5-HTP for a period of six weeks (Poldinger, 1991). The first group of thirty-six patients received 100 mg of 5-HTP three times a day with meals, while the second group of thirty-three patients received 50 mg of Luvox three times a day. At the conclusion of the study, results of the Hamilton Depression Scale, a widely used diagnostic test, had significantly improved in both groups.

As to side effects, 40 percent of patients in the 5-HTP group and 55 percent of patients in the Luvox group reported adverse reactions. The most common side effects in the 5-HTP group were gastrointestinal symptoms, including nausea and heartburn, while in the Luvox group nausea and weakness were the most predominant. In both groups, most of the adverse reactions were reported in the first few days of treatment. Laboratory studies did not reveal any significant changes.

The researchers say,

> *In essence, the results of this comparative study indicate both oxitriptan (5-HTP) and fluvoxamine to be distinctly effective in the treatment of depression, and, moreover, to be equally so. Regarding tolerance and safety, however, 5-HTP proved superior to fluvoxamine as was apparent in a marked difference in the severity of untoward side effects between the two compounds. Our results strongly confirm the efficacy of 5-HTP as an antidepressant.*

> *With all due deference to scientific skepticism, the reluctance shown by some authors of recent textbooks on the subject and by others to concede 5-HTP no place among acknowledged pharmacotherapeutics routinely applied against depression does not seem warranted, neither on empirical nor on theoretical grounds.*

Can 5-HTP affect more than just serotonin?

You may think that 5-HTP, since it becomes serotonin, influences nothing but that specific neurotransmitter. However, 5-HTP's actions may be less specific than originally thought. Apparently, 5-HTP can enter certain brain cells and also influence the release of dopamine and norepinephrine (Van Praag, 1984). In some laboratory studies on animals and humans, scientists have found these chemicals to be released into the synaptic cleft after 5-HTP is taken (Lichensteiger, 1967; Van Praag, 1983). While personally experimenting with 5-HTP, I've noticed a slight increase in visual and auditory clarity, which could perhaps be explained by increases in levels of norepinephrine and dopamine. 5-HTP does not convert into these chemicals, but just influences brain cells that contain them.

The multisupplement approach to treating depression

Even though some studies have shown 5-HTP to be effective in treating depression by itself, the dosages required were often high and led to side effects of nausea and sometimes even vomiting. I am always concerned about using high dosages of any particular medicine, especially one that has not been thoroughly tested for its long-term effects. Therefore, it appears that, for the time being, we should not rely exclusively on 5-HTP, and we need to find a way to combine it with other natural supplements for a safer and more effective method of treating depression.

With the current availability of a number of natural supplements that influence mood—particularly the herb St. John's wort—it is now possible, in my opinion, to not have to rely on pharmaceutical medicines for the therapy of mild, and probably of moderate, depression. Perhaps some cases of severe depression could even respond to a suitable combination of natural nutrients and herbs.

There are a number of medical and hormonal causes of depression, thyroid disease being a good example. A full medical evaluation is a must before starting any antidepressant program. I provide such a program in Chapter 11.

5-HTP and St. John's wort

St. John's wort is an herbal antidepressant. Since there are quite a number of active compounds within St. John's wort, it has been difficult to study the precise way or ways in which this herb works to relieve depression. In one study, extracts of St. John's wort caused a 50 percent inhibition of serotonin reuptake by rat brain cells (Perovic, 1995). The blocking of norepinephrine reuptake is an additional possibility (Muller, 1996). It seems that some compounds within St. John's wort also have the ability to inhibit MAO, but very weakly (Thiede, 1994). The antidepressant effects of St. John's wort are likely due to a number of factors working simultaneously. (For more information, see *St. John's Wort: Nature's Feel-Good Herb* in Appendix B.)

I've personally tried the combination of St. John's wort and 5-HTP. One morning, I made a cup of tea with two tea bags of St. John's wort. By noon, I was clearly feeling the alertness and mood elevation from the tea. At about 2 P.M., I started feeling a little overstimulated and edgy. I knew that 5-HTP would provide a sense of relaxation, but I had never before mixed it with St. John's wort. I took 25 mg of 5-HTP and within an hour, I could tell a sense of peacefulness had come on.

I left the office to take a walk by the sea, and started to ponder about which brain chemicals St. John's wort might really be influencing. Obviously, it wasn't completely raising dopamine levels, because there was no immediate and obvious increase in sex drive while taking this herb. St. John's wort wasn't exclusively influencing serotonin either, because when I added the 5-HTP, I felt the additional onset of relaxation. I had noticed some restlessness on St. John's wort, and perhaps this could have been because the herb was raising norepinephrine levels. I thought that this herb most likely did work in multiple ways.

After taking St. John's wort and 5-HTP in combination for several days, and noticing the continued mood elevation and relaxation, it appeared to me that this combination would be suitable for those who have a mild to moderate depression associated with anxiety. Restlessness and anxiety are sometimes reported with the use of St. John's wort.

5-HTP can help take the edge off. See Chapter 10 for interviews with doctors who have prescribed this combination.

Additional supplements that influence mood

There are a number of supplements that improve energy and mood.

Nutrients

Vitamins, amino acids, and other nutrients can be used.

B-complex vitamins—These vitamins have a mild antidepressant effect. They are involved in energy production, and are cofactors in the formation of neurotransmitters. As a rule, taking an amount anywhere up to five times the Recommended Daily Allowance (RDA) should be sufficient. For instance, the RDA for thiamine (vitamin B_1) is about 2 mg. Therefore, any dose of a B-complex that includes from 2 to 10 mg of vitamin B_1 should be fine.

When taking a B-complex, it wouldn't hurt to take a multivitamin and multimineral complex. For instance, the pill could include vitamin C, in amounts ranging from 100 to 400 mg, and vitamin E, in amounts ranging from 50 to 200 international units (IUs).

Tyrosine—This amino acid is used by the brain to create both dopamine and norepinephrine. It causes alertness and hence can counteract 5-HTP's sedative properties. Tyrosine is sold in 500-mg capsules, but your initial dosage should not be greater than 100 mg, especially if you are already on other nutrients. Simply open the capsule and divide the contents. In high dosages, this amino acid causes anxiety, restlessness, and insomnia; raises blood pressure; increases heart rate; and even causes heart irregularities in susceptible individuals. Phenylalanine is an amino acid that is converted into tyrosine within the body. The D,L form of phenylalanine is a good option.

Coenzyme Q$_{10}$—This antioxidant nutrient also increases energy levels. It is helpful in treating depressed individuals who have cardiovascular conditions such as congestive heart failure and high blood pressure. The daily dosage can range from 10 to 30 mg.

Lipoic acid—This powerful antioxidant provides a calm sense of well-being with a slightly enhanced sense of visual perception. One unique property of this nutrient is its ability to help regulate blood-sugar levels. Therefore, it is ideal for diabetics who also suffer from depression. The daily dosage can range from 10 to 30 mg.

Hormones

DHEA and pregnenolone, two hormones I have used both clinically and personally, are helpful for some people who have depression.

DHEA—This hormone has been shown in some studies, as well as in anecdotal experience, to have mood-elevating properties. Anecdotal evidence also indicates this hormone boosts sex drive because of its conversion into testosterone. Levels of this adrenal hormone decline with age, so DHEA is generally recommended only for individuals in their midforties and older. The maximum daily dosage for long-term use should not exceed 5 mg. High dosages can induce acne, hair loss, and even heart palpitations. (See *DHEA: A Practical Guide* in Appendix B, or my website, www.raysahelian.com, for more information.)

Pregnenolone—This hormone is the mother of DHEA and I call it the "Grandmother of all the adrenal hormones." It has properties similar to those of DHEA. Users notice enhanced mood and alertness, along with visual and auditory enhancement. The side effect profile is similar to that of DHEA. Regular daily dosage should not exceed 5 mg. (See *Pregnenolone: Nature's Feel Good Hormone* in Appendix B for more information.)

Herbs

There are at least three herbs besides St. John's wort that are helpful in influencing mood.

Kava—This herb is mostly used for anxiety, but also has slight mood-elevating properties. A dose of from 70 to 100 mg of the kavalactones—the chemical ingredient used to standardize kava extracts—is often enough to provide relaxation, mental alertness, and a sense of peacefulness. Kava can be taken any time of day, but late afternoon or early evening is the best option. Kava is best reserved for those in whom depression is associated with anxiety.

Ginseng—This herb has been used for centuries as an energy-improving herb. Quite a number of varieties are available in countless dosages; you can discuss dosage with an herbalist if your regular health care provider is not familiar with herbal medicine.

Ginkgo biloba—This herb is generally considered to be a memory booster, but it does have very mild alertness and mood-elevating properties. One 40-mg pill a day should be sufficient since, in this case, it is being combined with other supplements.

Why not take SSRIs instead of 5-HTP?

As I've said, these medicines elevate serotonin levels by preventing its reuptake, as opposed to 5-HTP, which provides the precursor to serotonin. Unlike SSRIs, 5-HTP is available without a prescription, and is a normal constituent of brain chemistry. The brain is accustomed to 5-HTP, while SSRIs are foreign drugs.

SSRIs have some unpleasant side effects, which include nausea, insomnia (mostly with Prozac), and loss of libido. Although 5-HTP has similar side effects, they are often less severe than those caused by the SSRIs. However, 5-HTP cannot always be substituted for these drugs. It is more appropriate for mild cases of depression, while SSRIs are more appropriate for moderate to severe cases. And while some patients prefer using natural supplements, others don't mind taking pharmaceutical medicines. It's even possible that we'll someday learn how to combine 5-HTP with these drugs. It is best to have an open mind and take advantage of all options when treating a particular medical or psychiatric condition.

One of the major advantages the SSRIs have over 5-HTP is that they have been studied much more exhaustively. This is, of course, a consequence of the enormous budgets pharmaceutical companies have for research and development when compared with the funds available for 5-HTP, which, because it is a naturally occurring substance, cannot be patented.

Using 5-HTP and SSRIs together

Studies using 5-HTP combined with various antidepressants are limited. Based on research results so far, though, it is possible that such combinations—if done appropriately and cautiously—could provide benefits.

Except for mild cases of depression, it is unlikely that 5-HTP, by itself, can be as effective as the SSRIs. If you have a mild to moderate case of depression, and wish to reduce or eliminate your dose of the SSRIs, you may need to use the antidepressive herb St. John's wort in addition to 5-HTP. Anytime you plan to switch medicines, do so only under full medical supervision. The change from medicines to supplements has to be done gradually.

It is well known that the antidepressants of the SSRI class, such as Prozac and Zoloft, elevate serotonin levels in the brain and body. Therefore, anyone who is currently on these medicines should be very careful in adding 5-HTP, since the effects increase each other. There is a good possibility that adding 5-HTP would decrease the required SSRI antidepressant dosage. When deciding to add 5-HTP, a doctor could start with 10 mg, and gradually, over several days, increase the dosage up to 50 mg, while gradually decreasing the SSRI dosage.

Another situation where 5-HTP could be helpful is in a case where a particular dosage of an SSRI is not fully effective and a doctor decides a higher dosage would be necessary. Instead of increasing the amount of the SSRI—for example, increasing the daily Prozac dose from 20 to 40 mg—it would be justified to add a small amount of 5-HTP daily and observe the effects for a few days. Perhaps the addition of the 5-HTP would eliminate the need for more Prozac.

Keep in mind that not all the SSRIs have the same actions. For instance, Paxil is known to be the most sedating, while Prozac is the least sedating. Also remember that Prozac has an extended half-life, which means its effects can last several days or more after being discontinued (Stokes, 1993). See Chapters 11 and 12, and talk to your health care provider about combining 5-HTP and SSRIs.

Using 5-HTP and the MAOI deprenyl together

As I have previously mentioned, serotonin is not the only brain chemical that influences mood. Another neurotransmitter that has a significant influence on alertness and well-being is dopamine. There are several medicines that are known to increase levels of dopamine, one of the most studied being deprenyl, also called selegiline.

In a double-blind study, eighteen patients with depression were treated with the combination of deprenyl and 300 mg a day of 5-HTP, twenty-one patients received 5-HTP alone, and another nineteen patients received placebo pills (Mendlewicz, 1980). The period of treatment lasted thirty-two days. Therapy with 5-HTP alone did not produce a significant improvement, but the combination of deprenyl and 5-HTP was very effective.

Deprenyl and 5-HTP may be a very good combination in cases of moderate or severe depression. One advantage of deprenyl is that it can increase sex drive, thus counteracting the lower sex drive that may be induced by 5-HTP.

In treating depression, a cautious starting dosage of 5-HTP would be 25 mg an hour before bed, and between 1 to 5 mg of deprenyl in the morning. Under medical guidance, these dosages could gradually be increased as needed. Daytime 5-HTP could be added, since the arousal from the deprenyl would counteract any possible sedation from the 5-HTP. See Chapters 11 and 12, and talk to your health care provider about combining deprenyl and 5-HTP.

Can 5-HTP change your personality?

Probably. A study published in the March, 1998 issue of the *American Journal of Psychiatry* indicates that antidepressants are able to change a person's personality, even in people who are not depressed (Knutson, 1998).

Drs. Kutson, Reus, and colleagues at the University of California—San Francisco report that a small group of nondepressed people who took Paxil became more easygoing and cooperative. "Different aspects of normal personality may be altered by psychopharmaceuticals that act on distinct nerve pathways in the brain," Reus said in a statement.

Twenty-three mentally healthy men and women took Paxil for a month, and filled out personality questionnaires before, during, and after. After just a week on the drug, they scored lower on tests that measured hostility and showed more cooperative behavior in puzzle-solving tests. A matched group of volunteers who took a dummy pill showed no changes. Side effects included daytime sleepiness and lowered sex drive.

5-HTP, from my clinical experience, works in a similar way to Paxil. Users report being more agreeable and relaxed. A number of other supplements influence mood and behavior, and hence personality, including St. John's wort, kava, and the hormones DHEA and pregnenolone.

Summary

5-HTP can be used for the treatment of mild depression, especially depression associated with anxiety. However, in cases of moderate to severe depressive mood disorders, it appears that 5-HTP cannot be relied on exclusively. The intelligent combination of 5-HTP with natural nutrients, herbs, and hormones can have powerful mood-elevating effects, eliminating the need for pharmaceutical medicines in mild cases of depression and, perhaps, even in moderate ones. At the least, the use of nutrients could allow for the reductions in the dosages of pharmaceutical medicines.

Since we don't know the long-term consequences of 5-HTP use, minimizing the dosage would be a cautious way to deal with this nutrient until more extensive studies are published. I would also recommend limiting the continuous use of 5-HTP to no more than three months, and cycling on and off the nutrient in order to avoid tolerance. There is a possibility that the effectiveness of 5-HTP may lessen with time (Van Praag, 1983). This would mean taking breaks from 5-HTP and, instead, using other nutrients and medicines for one or more weeks.

CHAPTER 6

A NATURAL OPTION FOR ANXIETY

Even during my very first few days of self-experimentation with 5-HTP, it was quite apparent to me that this serotonin precursor had marked calming properties. If the right dose of 5-HTP is taken, a feeling of peacefulness and steadiness comes on, without excessive sedation. After experiencing these feelings, I couldn't help but consider that 5-HTP had a role to play in anxiety disorders.

At about this time, I was starting to examine the stacks of scientific papers I had gathered on three decades of 5-HTP research. Although studies on 5-HTP and anxiety disorders are very limited, my personal experience and that of other doctors leads me to believe that this nutrient may play a role in reducing anxiety.

Understanding anxiety and its causes

The psychological state or symptoms of anxiety include worry and hypervigilance, which can be accompanied by mild depression. Some of the physical symptoms include muscle tension, increased heart rate, and increased blood pressure.

Stress and anxiety can be beneficial in terms of stimulating our brains and bodies to improve performance. However, when stress and anxiety are excessive, they can certainly impair normal functioning. Both social and occupational performance can be affected.

Some medical conditions can cause anxiety, and these must be ruled out before a diagnosis of anxiety can be made. Therefore, if the symptoms of anxiety are persistent, and are interfering with day-to-day activities, an individual would benefit from a complete medical evaluation. The health care provider will want to make sure the person does not have angina, in which a coronary-artery spasm causes chest pain; hyperthyroidism, in which the thyroid gland becomes overactive; or hypoglycemia, in which blood-sugar levels are too low.

A full medical evaluation is also important to rule out other physical causes of anxiety before taking any type of medication, and to ensure that the individual isn't taking substances that can, in themselves, cause anxiety. An evaluation for suspected anxiety disorder would include inquiries about usage of caffeine, alcohol, and over-the-counter stimulants and supplements. Certain cold medicines contain stimulants, such as pseudoephedrine. Even certain herbs can induce a state of nervousness, including ma huang, or ephedra, and high doses of St. John's wort. Amino acids, such as tyrosine and phenylalanine, increase energy and mood; however, nervousness, irritability, and heart palpitations are possible symptoms with higher dosages. High dosages of hormones, such as pregnenolone and DHEA, can induce irritability and anxiety in certain individuals. (For more information about these herbs and supplements, see Chapters 4 and 5.)

There are several types of anxiety disorders, including adjustment disorder, generalized anxiety disorder, obsessive-compulsive disorder (OCD), and panic disorder (PD).

Adjustment disorder with anxiety

This is the most common form of anxiety disorder. Almost everybody has experienced a stressful life event, such as a death or severe illness in the family, divorce, financial difficulties, and so on. The immediate consequence of this stress is anxiety. As a rule, this type of anxiety is short-lived, often lasting no longer than a few months, and medicines are not needed to treat it. Counseling, having open talks with friends or family, and relaxation techniques are adequate ways to treat this condition. However, if the anxiety persists, is causing enduring problems with sleep (see Chapter 7), or is interfering with work performance, individuals often visit their doctors for help.

Generalized anxiety disorder

This condition is often chronic, lasting six months or longer, and is associated with constant anxiety and worry. At least 4 percent of all patients who visit their primary care doctors suffer from this condition. Most symptoms begin during one's teens or twenties, often as a consequence of a particular stress, and the symptoms persist for years. These symptoms fluctuate depending on the amount of stress the patient is under. Anxiety is often accompanied by mild to moderate depression and alcohol overuse.

Since generalized anxiety disorder lasts for many years, an individual with a moderate to severe case may have to rely on medicines for a long time. The medicines most often prescribed by doctors are the benzodiazepines, such as Xanax (alprazolam) and Valium (diazepam), or newer medicines, such as Buspar (buspirone). These drugs do work, but there are drawbacks. Benzodiazepines are sedating, and also interfere with mental clarity. Buspar does not have a strong sedating property. However, it lacks the muscle-relaxing abilities of the benzodiazepine class. According to my discussions with several psychiatrists, Buspar is not always effective in relieving anxiety.

Another problem is the addiction potential that exists with many of these medicines, which leads to the medical and psychological problems associated with addiction. If 5-HTP is found to be helpful in partially or mostly relieving generalized anxiety, it could serve as a substitute for these drugs.

A study was done by doctors at the Department of Biological Psychiatry, State University in Utrecht, The Netherlands (Kahn, 1985). Ten outpatients with panic attacks, agoraphobia (fear of open spaces), or generalized anxiety disorder that had lasted more than one year were given 20 mg of 5-HTP a day, a dosage that was gradually increased as needed. The study was open, uncontrolled, and lasted for twelve weeks. The average dosage required for most of the patients turned out to be between 75 and 100 mg of 5-HTP per day. Nine out of the ten patients improved significantly on this dosage. The researchers state,

> *In view of the small number of patients, any conclusion must remain tentative. The present data, however, are encouraging and warrant further studies of the effect of 5-HTP in anxiety disorders.*

Obsessive-compulsive disorder (OCD)

Do you know someone who washes his or her hands more often than necessary? A friend who keeps an excessively clean house? Someone who arranges the closets very neatly? We certainly wouldn't consider any of these individuals to have a major problem. However, when such behavior is taken a step further, and goes beyond what we consider normal in our society, it's time to call it a disorder—OCD. This is especially true when the symptoms are associated with anxiety, depression, or hostility. For instance, if some-one washes his or her hands dozens of times a day, dusts the living room every hour, or keeps polishing the coffee table over and over again, that person has a problem.

Obsessions are defined as recurrent and persistent thoughts, ideas, impulses, and images that are intrusive and senseless. Compulsions are repetitive and intentional behaviors that are performed in response to the obsessions. These obsessions and compulsions cause marked distress, are time-consuming, and interfere with the normal func-tioning of a person's day-to-day life.

Serotonin is involved in OCD. A double-blind, multi-center study done by Dr. John Walkup of the Department of Psychiatry and Behavioral Sciences at Johns Hopkins University in Baltimore, Maryland, indicates that Luvox (fluvoxamine) is an effective short-term treatment for OCD in children and teenagers (*Clinical Psychiatry News*, p. 8, November 1996). A two-year study with adult OCD patients showed Zoloft (sertraline) to be effective without any sig-nificant adverse reactions or laboratory abnormalities (Rasmussen, 1997). Both Luvox and Zoloft are SSRIs, or antidepressants that increase serotonin levels within the brain. (For more information on the SSRIs and other anti-depressant drugs, see Chapter 5.)

The role of serotonin in OCD had been suspected for a long time. Back in 1977, researchers from North Nassau Mental Health Center in Manhasset, New York, gave seven OCD patients between 3 and 9 grams of L-tryptophan daily in divided dosages (Yaryura-Tobias, 1977). As we've seen, tryptophan is converted into 5-HTP within the body, which in turn is converted into serotonin. The patients also received B vitamins consisting of nicotinic acid, 1,000 mg twice a day, and pyridoxine, 200 mg twice a day. The nico-tinic acid was given because tryptophan can be metabo-lized into this vitamin. By providing high doses of nicotinic acid, more of the tryptophan would be available for 5-HTP creation. After one month, the patients showed significant improvement, and after six months to one year, their conditions stabilized. No major side effects were noted.

Drs. P. Blier and R. Bergeron, from the Neurobiological Psychiatry Unit, McGill University in Montreal, wanted to find out whether using 5-HTP and SSRIs together could help OCD patients who were resistant to the antidepressant drugs alone. Thirteen patients with OCD who had not responded to SSRIs were given 5-HTP without a noticeable improvement in symptoms (Blier, 1996). Interestingly, the use of tryptophan, the amino acid from which 5-HTP is created, with SSRIs produced a significant improvement after only four weeks. The reason for tryptophan's effectiveness and 5-HTP's ineffectiveness is not known. More studies are needed before we can come to any conclusions regarding the role of 5-HTP in OCD.

Panic disorder (PD)

This illness is marked by short-lived, recurrent, unpredictable episodes of intense anxiety called *panic attacks*. In addition to the psychological symptoms, there are physiological manifestations, including shortness of breath, rapid heart rate, headaches, dizziness, numbness, and nausea. Panic attacks can affect people with other forms of anxiety disorder. The difference is that the attacks associated with PD are spontaneous—they do not arise from specific situations.

PD generally affects young people. Approximately 1 to 3 percent of the population is affected at any one time, and the female-to-male ratio is 2 to 1. The attacks occur more commonly in the premenstrual period (see Chapter 9) and during times of stress.

Several studies have shown that SSRIs, such as Prozac and Paxil, are effective in the short-term treatment of PD

(Den Boer, 1990; Ballenger, 1998). However, there are patients who do not respond to SSRIs, and some even find that their symptoms worsen. At this point, we don't know whether the regular use of 5-HTP helps patients with PD, but there's a theoretical possibility that it would. A few patients with mild cases of PD have noticed some benefit. Shirley, a 27-year-old actress from Hollywood, says,

> *I have panic disorder that manifests itself in the usual way: I can't breathe, have chest discomfort, and then I just start crying. Although I take Trazodone and Paxil [both antidepressants] to keep the edge off, I have found something that really works on those days where I just can't stand to stay at work for one second more or I'll die. 5-HTP really works to the point where time goes much faster for me when I'm somewhere I don't want to be.*

No formal studies have been done to test 5-HTP as a therapy for PD. However, doctors have given this supplement in short-term dosages to PD patients in order to test the patients' hormonal responses. Let's discuss a few of these studies.

Researchers at the Department of Biological Psychiatry, Academic Hospital in Utrecht, The Netherlands, gave 5-HTP to seven female patients with PD and seven healthy volunteers (Westenberg, 1989). The single dose used was 60 mg, given intravenously. None of the PD patients reported a panic attack or an increase in anxiety. In contrast to the expectations of the researchers, the patients felt more relaxed and less anxious, starting about one hour after the infusion was completed. The most common side effects reported were nausea and drowsiness. Other reported effects were mood elevation and difficulty in concentration. Two patients mentioned feelings of unreality.

In 1990, the same team of investigators in The Netherlands who did the 1989 study gave twenty PD patients single 60-mg doses of intravenous 5-HTP. The researchers compared the results with those found in a group of twenty healthy volunteers (Den Boer, 1990). During and after the 5-HTP infusion, none of the PD patients showed an increase in anxiety or depressive symptoms, despite the presence of severe side effects such as nausea and vomiting. In both the patients and the controls, the 5-HTP infusion led to substantial increases in blood levels of two hormones, cortisol and beta-endorphin, involved in mood control. In addition, levels of melatonin—the master sleep hormone that also elevates mood—increased significantly.

Dr. Irene van Vliet and colleagues, also from Academic Hospital in Utrecht, gave 5-HTP to seven patients with PD and seven healthy volunteers (van Vliet, 1996). Intravenous doses of 10 mg, 20 mg, or 40 mg, or a placebo, was given on four different occasions. Blood levels of 5-HTP, cortisol, and 5-HIAA, the breakdown product of serotonin, were measured at regular intervals. None of the patients or controls experienced panic attacks, or showed increases in levels of anxiety or depression. There were side effects, such as nausea, dizziness, and fatigue. The higher the dose, the more frequently the side effects occurred.

Only the 40-mg infusion of 5-HTP led to increases in blood levels of cortisol. During the infusion, there were lowerings of heart rates and systolic blood pressures. (For more information on 5-HTP and its influence on hormones, see Appendix A.)

Other supplements that promote calmness

I believe that the natural approach to treating anxiety should employ several supplements working together. Kava, a South Pacific herb, is especially useful for anxiety accompanied by depression. A good dosage is between 70 and 100 mg of the kavalactones, three times a day. The mineral magnesium and the amino acid taurine can also play a role. In addition, a number of sedative herbs may be helpful, including valerian, hops, passionflower, and chamomile (see Chapter 7). In Chapter 11, I will provide practical guidelines on how to use 5-HTP in conjunction with other relaxing herbs and nutrients. I would strongly advise you to see your health care provider before starting any antianxiety program, especially if you are currently on prescription medication.

Summary

Serotonin is a brain chemical involved in both relaxation and mood stabilization. Since serotonin is created from 5-HTP, it makes sense that this nutrient would promote feelings of peacefulness and calmness. This serotonin precursor should definitely be considered as an additional supplement in the therapy of anxiety disorders. Of course, before taking any medication, all other treatments should be attempted. This includes counseling, cognitive-behavioral therapies, muscle relaxation techniques, relaxation books and tapes, guided imagery, yoga, breathing exercises, meditation, sincere talks with close friends and family, and breaks from the patient's standard routine.

Since the long-term use of any antianxiety medicine could have harmful physiological effects, one option for someone who requires such medicines for many years would be to take a particular medicine for a few weeks or months and then slowly switch to another. This way, the body will not be exposed to the same type of drug, thus reducing the potential harm or damage to tissues and organs. The possibility of developing addiction and tolerance would also be reduced.

If 5-HTP and kava, along with other herbs and nutrients, can be effectively used for a few weeks or months to reduce anxiety and thus eliminate or decrease the need for pharmaceutical drugs, then such a program would do a great service to many patients. Another option would be to alternate the use of several medicines and nutrients on a daily or weekly basis.

CHAPTER 7

TREATING INSOMNIA THE NATURAL WAY

Everyone has a restless night's sleep from time to time, a situation that does not require medicine. Chronic insomnia, though, can cause a person to become irritable and fatigued, and lose the ability to concentrate, which in turn can affect relationships at home and at work. It can also interfere with optimum health.

Tryptophan—the amino acid used by the body to create 5-HTP—was prescribed in the 1980s for insomnia, since studies had reported that it increased both total sleep time and the time spent in deep sleep (Hartmann, 1987). For many people, 5-HTP is also effective in inducing sleep. My personal clinical experience in this matter has been confirmed by my discussions with doctors and by published research (Soulairac, 1990). Within a half-hour of taking this serotonin precursor in the late evening, most people will notice the agreeable onset of a yawn, followed by the urge to go to bed. This urge can be resisted if necessary. It's not like taking a prescription sleeping pill that would knock you out.

However, 5-HTP does have a couple of shortcomings. It is not consistent in inducing and maintaining sleep, and it can enhance vivid dreaming. These shortcomings can be offset by using this nutrient as part of an overall program for insomnia relief.

Tryptophan, 5-HTP, serotonin, and sleep

How 5-HTP works to induce sleep is not fully understood. However, the explanation could be as simple as the nutrient's conversion into serotonin, which itself has sedative properties, and subsequently into melatonin, the body's primary sleep hormone (see Chapter 3). Conversion into melatonin occurs in the pineal gland, the pea-sized gland located in the middle of the brain that is responsible for helping us sleep at night. Dr. Al Lewy, a melatonin researcher from the Sleep and Mood Disorders Laboratory at Oregon Health Sciences University in Portland, confirms that no other area in the brain is currently known to make melatonin except for the pineal gland (personal communication).

Doctors suspect that melatonin is also made in the gastrointestinal system (personal communication with Dr. Gerald Huether, Psychiatric Clinic, University of Göttingen, Germany). Therefore, we should consider the possibility that 5-HTP could be converted into serotonin and then on into melatonin in the intestines, thus increasing circulating melatonin levels in the blood. This possibility has not been fully evaluated.

The polytherapy approach to chronic insomnia

There are several supplements available over the counter that can induce sedation and sleep. There are also numerous sleeping pills commonly prescribed by physicians for those with insomnia. Often, people take one particular pill on a continuous basis for months or years. My concern with this *monotherapy* approach is the fact that the brain could build a tolerance to one particular pill, and the chronic use of one chemical could have harmful effects.

For these reasons, I prefer the *polytherapy*, or multisupplement, approach to treating insomnia. By alternating different natural supplements, and even prescription pills when needed, one would run a lower risk of tolerance, and, one hopes, a lower risk of side effects. It's also possible that combining small amounts of more than one drug or supplement would strengthen the effect of each, since each would work in a different way. Of course, such an approach should be considered only after a full physical examination is performed, in order to rule out the possibility that a physical disorder is causing the insomnia. A step-by-step guide to natural sleep therapy is provided in Chapter 11.

Additionally, it's almost impossible to sleep well every night if you don't engage in some sort of physical activity. Your body must be tired when you go to bed. A good time to be physically active is from three to six hours before bedtime. During exercise, your body will become warmer, and over the next couple of hours it will cool off, allowing for better sleep.

Let's briefly discuss a few sleep-inducing therapies.

Melatonin and 5-HTP

Melatonin is what I call "Nature's sleeping pill." It's probably the most consistent natural over-the-counter sleep aid available, since it's the actual hormone our brains release every night to help us sleep. Over the past three years, I have taken from 0.3 to 1 mg of melatonin about once or twice a week, so I know the effects of this hormone very well. I recommend it in a range of dosages and also in a variety of forms—regular pills, sublingual lozenges (which are placed under the tongue), time-release capsules, liquid extract, and even tea. Melatonin works well for most patients, but tolerance can develop when it's used every night. Side effects on dosages greater than 1 mg include vivid dreaming and morning grogginess. (For more information, see *Melatonin: Nature's Sleeping Pill* in Appendix B.)

Since 5-HTP converts into melatonin, why not just take melatonin and not bother with 5-HTP? The answer to this question is still the subject of much research. Having tried both substances, I find them to be similar in their effects. At this time, all I can say is that individual responses to each supplement vary, and some people respond better to one than to the other. Both 5-HTP and melatonin induce and maintain sleep, but they are not always effective, and there are individuals who seem to not respond to these supplements at all.

Both melatonin and 5-HTP, when used every night for a couple of weeks or longer, can produce middle of the night, or early morning, awakening. For this reason, neither of them should be used on a nightly basis. The mechanism for this action is not clear. Perhaps there's a tolerance that develops to these substances, or the body learns to break them down more quickly, or the brain cell receptors that respond to them decrease in number.

Dreams are enhanced on both 5-HTP and melatonin. Vividness depends on dosage: the higher the dosage, the more vivid the dream. Depending on the content, a vivid dream can be very enjoyable, providing many insights into one's subconscious, or it can turn out to be extremely frightening.

Herbs and herbal combinations

There are a number of over-the-counter products that combine several herbs, such as valerian, kava, hops, passionflower, chamomile, and others. These combinations are also available in tea form. You can certainly drink these teas one or two nights a week if you have insomnia. If you use single herbs, I would suggest taking them in rotation, rather than taking any one herb repeatedly, in order to avoid tolerance.

Valerian—This herb has been used since the Middle Ages for nervous disturbances affecting the gastrointestinal system, and to induce sleep (Holzl, 1989). Preliminary research indicates that certain compounds in valerian influence some of the same brain receptors affected by the antianxiety medicines in the benzodiazepine class, such as Valium and Xanax (Holzl, 1989); see Chapter 6 for more information about these drugs. These receptors are activated by gamma-amino-butyric acid (GABA), an amino acid that functions as a neuotransmitter. Activation of GABA receptors leads to relaxation and sleep (see page 79).

Valerian is not nearly as powerful and consistent in inducing sleep as are 5-HTP and melatonin. However, a study of fourteen elderly persons who slept poorly showed that 400 mg of a valerian preparation, taken three times a day for one week, decreased the amount of time it took the subjects to fall asleep. It also enhanced deep, slow-brain-wave sleep (Schulz, 1994). Rapid eye movement sleep, in which most dreaming occurs, was not affected. This study needs to be interpreted cautiously, since the number of subjects was small.

Since 5-HTP induces sedation and sleep by influencing the body's serotonin system, and by its ultimate conversion into melatonin, it appears that this nutrient would combine well with valerian, which influences the GABA system. Valerian does not seem to induce vivid dreaming, nor does its use in proper dosages lead to any significant morning hangover.

Generally, a dose of from 200 to 500 mg of concentrated valerian root extract is effective in inducing and maintaining sleep. The extract should provide from 0.5 to 1 percent of the essential oils, such as valerinic acid. Take the herb anytime from one-half to three hours before retiring. The tea form also works well. Again, as with both 5-HTP and melatonin, the results with valerian are not always consistent, and some individuals have the paradoxical response of becoming more alert.

Kava—This is a South Pacific herb that contains kavalactones, chemicals that reduce anxiety. The ingestion of kava induces alertness with relaxation, and after a few hours, there's a feeling of sedation. Some individuals do find that a small dose of kava promotes sleep. Kava should not be taken too close to bedtime, though, since, in some people, it may interfere with sleep. It can be taken in late afternoon or early evening, at a dose of 40 to 70 mg of the kavalactones. (For more information, see *Kava: The Miracle Antianxiety Herb* in Appendix B.)

Hops—For the past millennium, hops has been used in the brewing of beer. In addition to compounds called flavonoids, a number of resins are found in hops, including humulone, lupulin, and colupulone.

This herb has sedative qualities, which, in my experience, are stronger than chamomile and passionflower, and somewhat similar in strength to valerian. In fact, hops and valerian are used to ease nervous conditions. Hops can be used by itself at night to induce sleep, or it can be combined with other sedative herbs. It can be taken in tea or capsule form.

Passionflower—This herb acts gently to promote sedation, but its effects are very weak. Passionflower is available in both liquid extract and capsule form.

Chamomile—This is one of the best-known herbal sleep aids. However, its effects are extremely weak. It may slightly ease digestive problems associated with stress. The bagged tea is available commercially, as are chamomile oil, capsules, and tincture. There are occasional reactions to chamomile in those who are allergic to ragweed.

GABA

GABA (gamma-amino-butyric acid), an amino acid found almost exclusively in the brain, serves as a neurotransmitter that induces relaxation. Disorders in GABA processing are suspected to play a role in a number of neurological conditions, including Huntington's disease, Parkinson's disease, seizure disorders, and schizophrenia (Cooper, 1996).

GABA is available without a prescription. However, trying to raise brain GABA levels by taking a pill is difficult, since this amino acid does not easily cross the blood-brain barrier. I have taken GABA on several occasions in doses ranging from 500 to 2,500 mg. Except for a temporary feeling of relaxation, I did not notice much of an effect.

Diet and sleep

You may be surprised to learn that the types of food one eats in the evening will significantly influence whether one is alert or sleepy (see Chapter 3). Therefore, if you wish to induce sleep at night, I recommend you eat a small to moderate meal in the late evening consisting mostly of carbohydrates. These include grains, legumes, pasta, bread, vegetables, fruits, and cereals. You can also have a small amount of fat or protein. For instance, if you plan to eat a whole grain cereal, you can add some milk to it. However, the bulk of the meal should consist of carbohydrates.

Prescription sleeping pills

I have no objections to the occasional—for instance, two to four times a month—use of a pharmaceutical sleep aid by individuals who have moderate to severe cases of insomnia and do not seem to respond to natural therapies alone. I'm not a purist in the sense that I think natural supplements should be used exclusively for every disorder. There are many situations in which pharmaceutical medicines can be of tremendous benefit if used appropriately.

There are a number of prescription drug options available. These include Valium (diazepam), Xanax (alprazolam), Ativan (lorazepam), Dalmane (flurazepam), Restoril (temazepam), and Ambien (zolpidem). Talk to your health care provider about which medication would be most helpful in your case.

Insomnia in children

There are certain children who have a long sleep-wake cycle. I had this problem when I was a kid: I could never immediately turn off my mind and relax my body when I was put to bed. It would take me at least a half-hour to an hour of tossing and turning before I finally got to sleep. I now suspect that my sleep-wake cycle probably did not follow the twenty-four-hour earth cycle, but was extended to maybe twenty-five hours or longer. In other words, if I was sleepy at 10 P.M. one night, the following night I wouldn't be sleepy until 11 P.M., and the night after, not until midnight. I could stay up very late and had a tendency to sleep in late, too.

I've recommended the polytherapy insomnia approach to parents with children who have difficulty sleeping, and it has worked well. Theresa, a mother from Long Beach, California, says,

> My six-year-old son has had trouble sleeping ever since I can remember. We put him to bed at 10 P.M. and we're lucky if he falls asleep by midnight. I followed your suggestions three months ago on using melatonin

and we now give him a quarter of a 0.5 mg tablet twice a week. It has had an enormous influence on him. The teacher says he's more attentive in school and his performance is better. One month ago, we added 10 mg of 5-HTP and it's working well, too. We're only using 5-HTP once a week. I will occasionally give him teas of passionflower, valerian, chamomile, or hops. It's amazing how better behaved and relaxed he is now.

The dosages used in treating childhood insomnia are about a fifth to a third of the adult dosage. A health care provider should supervise children who are treated with these supplements.

Summary

Although many natural supplements and pharmaceutical medicines are effective sleep aids, the majority of these substances can induce tolerance when used every night. Hence, a wise approach is to take advantage of a variety of substances, use the smallest dosages that work, and alternate their use. This way, any possible risk from the use of one substance would be minimized. I also recommend not taking anything for sleep at least one or two nights a week.

The true promise of 5-HTP as a sleep aid lies in its occasional use as an additional option for individuals interested in taking a nonpharmaceutical approach to insomnia relief.

CHAPTER 8

EASING THE SYMPTOMS
OF FIBROMYALGIA

S o far, we've seen that 5-HTP plays a role in weight loss, depression, anxiety, and insomnia. Might it also ease some of the symptoms of the painful condition known as fibromyalgia? Also referred to as fibrositis, fibromyalgia is one of the most common rheumatic conditions in medicine, affecting about 2 to 5 percent of the general population (Wolfe, 1995).

5-HTP has shown some promise as natural therapy for this disruptive and occasionally disabling disease, especially when it is combined with other therapies. But first, let's take a closer look at fibromyalgia itself.

Fibromyalgia: that run-over feeling

Women between twenty and fifty years old are the most likely to experience the symptoms of fibromyalgia, which include fatigue, irritability, memory problems, headaches, and anxiety. But the most notable symptom is pain, especially that felt in *tender points*, or points sensitive to pressure, scattered throughout the body. These tender points tend to occur in specific areas in the neck, rib cage, back, thighs, buttocks, knees, and elbows. Muscles and other forms of soft tissue can ache. A common answer by fibromyalgia patients to the question, "How do you feel in the morning?" is what's being called the "18-wheeler" response, as in, "I feel like I was hit by a Mack truck." (Sigal, 1998)

Several medical conditions are often closely associated with fibromyalgia. These include (Wallace, 1997):

- Sleep disorders

- Irritable bowel syndrome

- Chronic fatigue syndrome

- Multiple chemical sensitivities

There is also a high prevalence of migraine headaches in patients with fibromyalgia, indicating that perhaps problems with serotonin, which is an important neurotransmitter, is one of the common denominators in both disorders.

No definite cause can be found for fibromyalgia and no laboratory tests are available to diagnose it, diagnosis being based on the nature of the symptoms themselves. Several factors are known to aggravate this condition, including tension, depression, excessive work activity, and changes in weather.

Patients suffering from fibromyalgia are often resistant to various forms of treatment. One of the standard medical therapies for fibromyalgia is the use of tricyclic antidepressants, such as amitriptyline, which block the reuptake of serotonin and other neurotransmitters. (See Chapters 3 and 5 for discussions on neurotransmitters and antidepressants, respectively.)

Research on 5-HTP and fibromyalgia

Success with the use of amitriptyline has prompted researchers to see whether providing serotonin precursors, such as 5-HTP, could relieve this condition. Another rationale for the use of 5-HTP is that low serotonin concentrations are associated with a reduced threshold of pain (Morgane, 1981).

In the last few years, at least three studies, all done in Italy, have evaluated the role of 5-HTP in fibromyalgia treatment.

The 1990 Milan study

Researchers at the Rheumatology Unit of Sacco Hospital in Milan evaluated patients with fibromyalgia in a double-blind, placebo-controlled trial (Caruso, 1990). "Double-blind" means that neither the researchers nor the patients knew who was taking the 5-HTP and who was taking the placebo. The patients were both male and female, and ranged in age from eighteen to sixty-five years. All of the patients had at least seven tender points. In addition, the patients had to have reported at least two of the following symptoms to be included in the study: muscle aches in various parts of the body, anxiety, morning stiffness, and irritable bowel syndrome.

Using random selection, twenty-five patients were given 100 mg of 5-HTP three times a day for thirty days. Another twenty-five patients received a placebo pill three times a day for thirty days. The placebo and 5-HTP pills were identical in color, shape, and size.

Treatment with 5-HTP resulted in a significant decline in the number of tender points and in the intensity of pain when compared with the placebo. A significant improvement was also reported in morning stiffness, sleep patterns, anxiety ratings, and fatigue ratings. No changes were noted in laboratory studies before and after treatment, including complete blood count, urinalysis, and blood chemistry profile.

Six patients out of twenty-five (24 percent) reported side effects with the 5-HTP, which included headache, diarrhea, and stomach pain. In the placebo group, three out of twenty-five (12 percent) reported side effects. The researchers state,

> The encouraging results observed in the present study may be due to the fact that orally administered 5-HTP is well absorbed by the peripheral tissues [and] the central nervous system by simple diffusion, with concentrations in the central nervous system being directly dependent on [blood] plasma 5-HTP concentrations. Interference by other amino acids is minimal, unlike the situation with [5-HTP's] precursor, tryptophan. It is concluded that 5-HTP is effective in improving the symptoms of primary fibromyalgia syndrome and that it maintains its efficacy throughout the 30-day period of treatment.

The 1992 Milan study

Caruso and his colleague, Puttini, conducted a study similar to the 1990 study, this time providing 100 mg of 5-HTP three times a day for a period of ninety days (Puttini, 1992). At least 50 percent of the patients improved. No major side effects were reported, and no unusual findings were observed in laboratory testing. The researchers state,

> 5-HTP confirms its efficacy in the treatment of primary fibromyalgia syndrome: the drug achieved a statistically significant effect after only 15 days of treatment and retained this effect for up to 90 days of treatment.

The 1996 Florence study

This study was done at the Interuniversity Center of Neurochemistry in Florence. Four groups of patients with fibromyalgia and migraines were given either tricyclics, another class of antidepressants called monoamine oxidase inhibitors (MAOIs), 400 mg a day of 5-HTP alone, or 5-HTP with MAOIs. The combination of 200 mg of 5-HTP a day with the MAOIs resulted in significant improvements in fibromyalgia symptoms during the twelve-month study (Nicolodi, 1996). The benefits induced by this combination were much greater than those induced by any of the three substances alone. Common side effects caused by the combination included insomnia and nausea. (See Chapter 12 for more details.) Laboratory tests during the study period did not show any abnormalities.

Additional therapies to consider

Fibromyalgia is very difficult to treat, and no single therapy is likely to be completely effective. Some patients find benefit in improving their dietary habits. This includes decreasing saturated fat intake, increasing fish consumption, and eating a variety of fruits, vegetables, and grains. A basic multivitamin supplement could help, along with additional antioxidants such as vitamins C and E. Small amounts of the antioxidants coenzyme Q_{10} and lipoic acid, along with the mineral magnesium and the nutrient carnitine, could also be used. If the fibromyalgia is accompanied by depression, St. John's wort is a good option. Some patients find the occasional use of small doses of hormones such as DHEA or pregnenolone—up to 10 mg a day—to be helpful. It may take several trials with many types of nutrients to find the ones that provide the best results.

Of course, gentle exercises and walking are beneficial. Since many of the symptoms and disorders associated with fibromyalgia are stress-related, a stress-reduction program may help. Breathing exercises, meditation, biofeedback, and yoga are just some of the possibilities. See your health care provider for advice.

One of the major problems with fibromyalgia is the continuous absence of deep sleep. Fortunately, as we saw in Chapter 7, 5-HTP is an effective treatment for insomnia, especially when used as part of an overall insomnia-relief program.

Fibromyalgia: a personal story

Miryam Williamson, from Warwick, Massachusetts, has told me a fascinating story about her lifelong attempts to find help for her insomnia, aches, and pain. She says,

> *Ever since I can remember, back when I was five years old, I've had trouble sleeping. Aches and pains were a routine part of my life. It was normal for me to be in constant pain. I never knew what it was like to not be aching all over. I always knew there was something wrong with me.*
>
> *I went to many doctors and finally gave up on the medical profession. I did take tryptophan in the 1980s and it helped with sleep. I took tryptophan—1,500 to 2,000 mg each night—for at least four to five years without side effects. I couldn't sleep without it. When the FDA took it off the market in 1990, I still had a stash for six months, then it was all gone. Subsequently, I went a year and a half getting hardly any sleep.*
>
> *It wasn't until 1993, at the age of 57, that I finally went into a crisis and couldn't function. My husband insisted I visit a doctor. I lucked out: the doctor knew what I had and diagnosed me with fibromyalgia. Her first prescription was for amitriptyline (Elavil) at 10 mg at night. It helped a little bit, but I still didn't sleep very well. She upped the dose to 20 and then 30 mg. I still didn't like the way I felt. I slept some, my pain decreased, but I felt like a zombie most of the time.*

After the fibromyalgia diagnosis, I focused on finding a replacement for tryptophan. I finally came across information through an on-line search of the National Library of Medicine's Medline database on 5-HTP in late 1994. I gave the information to my doctor and she agreed to prescribe it for me through a compounding pharmacy. I started taking the 5-HTP beginning in March of 1995. I gradually lowered my amitriptyline dose by 10 mg each week, while I added 100 mg of the 5-HTP. I was off the amitriptyline in three weeks. 5-HTP allowed me to sleep well, be less fatigued during the day, and feel fewer aches and pains.

It's been now three years and I'm still on 300 mg of 5-HTP almost every night. If I don't take it, I wake up in two to three hours. I also take 1 mg of melatonin and 25 mg of [an antihistamine called] diphenhydramine.

For more information, see *Fibromyalgia: A Comprehensive Approach* and *From Fatigued to Fantastic!* in Appendix B.

Summary

Fibromyalgia is a chronic disorder that can last for years and decades. At this point, despite the anecdotal evidence of certain individuals, I would recommend the use of 5-HTP only two or three times a week, and a maximum dosage of from 50 to 100 mg, since we don't know the long-term effects of daily 5-HTP use. This nutrient should be used as part of an overall treatment program, especially in light of the multiple disorders that often accompany fibromyalgia.

CHAPTER 9

OTHER DISORDERS AND 5-HTP

As we've seen, 5-HTP's benefits result from its effects on serotonin production. The fact that serotonin has a number of functions within the body means that 5-HTP can potentially play a role in treatment plans for a number of different disorders. However, much of this research is in its earliest stages, and there is much that we must learn before all of 5-HTP's benefits and shortcomings are known.

5-HTP may play a part in the treatment of premenstrual syndrome, a number of psychiatric ailments, several neurological disorders, and certain childhood disorders.

5-HTP and premenstrual syndrome (PMS)

PMS—also known as late luteal phase dysphoric disorder (LLPDD) or premenstrual dysphoric disorder (PDD)—occurs, to some degree, in one-third of all women who are of childbearing age. The symptoms include irritability, increased aggressiveness, cravings for sweet or salty foods, nervousness, mood swings, difficulty in concentrating, depression, fatigue, breast tenderness, and abdominal bloating. These symptoms appear during the latter half of the menstrual cycle and disappear with the onset of menstruation.

The exact causes of PMS are not fully known, but hormonal imbalances or abnormal responses to fluctuating levels of hormones such as progesterone during the late part of the menstrual cycle are thought to be involved. These hormonal changes are associated with brain chemical changes.

Serotonin is suspected to play a role in PMS, since the use of antidepressants called selective serotonin reuptake inhibitors (SSRIs) can decrease symptoms. A study published in the *Journal of the American Medical Association* showed that Zoloft (sertraline) improved PMS symptoms in a group of two hundred women. Of those women who took the SSRI, 62 percent improved, versus 34 percent of the women who took a placebo (Yonkers, 1997). Another SSRI, Paxil (paroxetine), has shown similar benefits (Sundblad, 1997).

Hormonal therapy is an option in the treatment of PMS. Two hormones to consider are progesterone and pregnenolone. Synthetic progesterone is used in some birth control pills, but this hormone is also available in natural forms. Pregnenolone, which can be converted into progesterone within the body, can be taken for a few mornings, once symptoms begin, at a dosage of from 5 to 10 mg. (See *Pregnenolone: Nature's Feel Good Hormone* in Appendix B.)

5-HTP should certainly be tried by itself. But if it is not effective, the best approach might be to use a combination of treatments. Perhaps pregnenolone in the morning followed by 50 mg of 5-HTP at night would be a good combination. Another option is to use 300 mg of the antidepressant herb St. John's wort (see Chapter 5) in the morning followed by 5-HTP at night. One could also try a 300-mg pill of St. John's wort in the morning combined with from 5 to

10 mg of pregnenolone or progesterone, along with nighttime 5-HTP. These supplements should only be taken during the premenstrual phase. A B-complex vitamin in the morning would also be advised. Each woman is likely to have a unique response to these nutrients and hormones. Therefore, if you want to use these substances in PMS relief, I strongly recommend that you seek the guidance of a health care provider. Daily exercise can certainly be helpful.

5-HTP and psychiatric disorders

There is no doubt that serotonin is one of the most important neurotransmitters in the brain, and that medicines which manipulate serotonin levels will influence human behavior. Even the behavior of animals can be manipulated by giving them SSRIs. For example, researchers at the Department of Psychology, University of Massachusetts, gave Prozac (fluoxetine) to a group of prairie voles (Villaba, 1997). Both the treated voles and a group of untreated animals were exposed to intruders in their cages. The voles treated with Prozac showed more tolerance and less aggression towards the intruders. Given the great complexity of the human brain, it should not surprise us to realize there is much more to learn about the links between brain chemistry and human behavior.

Addiction disorders

There's been a search for decades to find a drug that could help alcoholics abstain from drinking. Several agents have been used, including Antabuse (disulfiram), an alcohol-sensitizing drug; antianxiety agents, such as Buspar (buspirone) and the benzodiazepines; and antidepressants.

It has been thought that the elevation in brain levels of dopamine and serotonin, two important chemicals involved with mood, could help indviduals with alcoholism avoid relapses. Treatment with Prozac, 20 mg a day for twelve weeks, showed this SSRI to be helpful in both improving depression and decreasing alcohol intake (Cornelius, 1997).

In order to evaluate whether 5-HTP could play a similar role, Dr. D. George and colleagues at the National Institute on Alcohol Abuse and Alcoholism in Bethesda, Maryland, gave 5-HTP and either carbidopa, levodopa and carbidopa, or a placebo to patients with alcoholism for one year (George, 1992). Carbidopa is a peripheral decarboxylase inhibitor (PDI), a substance that prevents 5-HTP from being converted into serotonin outside of the brain (see Chapter 11). Levodopa is known to elevate brain levels of dopamine, another important brain chemical. The 5-HTP was started at 25 mg four times a day—breakfast, lunch, dinner, and bedtime—and increased in some patients to 100 mg four times a day.

Only eight of the thirty-one patients who entered the study abstained from alcohol for the entire year. There was no significant effect of any treatment on the ability to maintain abstinence. The most common side effects were sleepiness and gastrointestinal disorders. The researchers conclude, "Long-term loading with precursors of dopamine and serotonin is probably not effective in improving the rate of abstinence in patients with chronic alcohol dependence."

One unfavorable study does not mean that 5-HTP cannot play a role in alcoholism treatment. For example, a trial combining 5-HTP with St. John's wort, B vitamins, or other nutrients would be a worthwhile project.

Another interesting area to explore is the use of 5-HTP, St. John's wort, and other mood-influencing nutrients in cases of drug and tobacco addiction, as well as in cases of pathological gambling. It has been speculated that many pathologic gamblers have an associated psychiatric disorder, be it depression or manic-depressive disorder. In a study reported in the October 1997 issue of *Clinical Psychiatry News*, Dr. Eric Hollander and Dr. Concetta DeCaria, of the Mount Sinai School of Medicine in New York, reported that seven of ten pathologic gamblers responded to treatment with the antidepressant Luvox (fluvoxamine). This would indicate that serotonin is involved in excessive and uncontrolled gambling.

Multiple chemical sensitivities (MCS)

This disorder is characterized by various symptoms following exposure to very low levels of common chemical substances. Patients with MCS have been considered to have unusual psychiatric disorders, or to suffer from anxiety or depression. A case study was published in a Scandinavian psychiatric journal of a person with MCS whose symptoms cleared after therapy with an SSRI (Ronnback, 1997). There may be subgroups of MCS patients who would respond positively to medicines that influence serotonin, and perhaps to 5-HTP as well.

Schizophrenia and other psychotic disorders

The schizophrenic disorders are a group of syndromes characterized by significant disruption of mood, thinking, and overall behavior. Alterations in brain chemistry are associated with the symptoms of this condition, and serotonin is one of the neurotransmitters thought to play a role.

A study performed at the Clinical Research Center on Alcoholism, Veterans Administration Medical Center in San Diego evaluated the role of 5-HTP in reducing amphetamine-induced aggravation of psychotic symptoms among schizophrenic patients (Irwin, 1987). Amphetamines are

known to release chemicals in the brain, such as norepinephrine and dopamine. This release can temporarily
aggravate psychotic symptoms in patients who have schizophrenia. By giving these patients 5-HTP before giving
them amphetamines, the scientists were able to significantly reduce thought disturbances and hallucinations. The
researchers state, "We believe that these data provide additional evidence for the concept that serotonin may be
important in the modulation of psychotic symptoms in
schizophrenic patients."

The role of 5-HTP in the therapy of schizophrenia has
not yet been fully evaluated. The results of preliminary trials have not been consistent (Wyatt, 1972; Bigelow, 1979),
and additional studies are needed.

Sexual disorders

As a rule, serotonin depletion increases sexual drive, while
serotonin excess decreases it. It is believed that a number of
sexual disorders could be treated with medicines that
increase serotonin levels. Theresa Crenshaw, M.D., in her
book *Sexual Pharmacology* (1996), writes, "Treatment of sexual behaviors, such as paraphilias [socially unacceptable
behaviors, such as exhibitionism], sexual aggression, and
various sexual compulsions, addictions and perversions,
may be possible with serotonin potentiators." 5-HTP has
not been formally tested in this regard, but it could theoretically be beneficial.

Another area in which the use of 5-HTP could be
explored is premature ejaculation. It has been shown,
through the use of SSRIs, that excessive serotonin can delay
orgasm. It is possible that 5-HTP would have the same
effect. This was true for one of my patients. Joe, a 35-year-
old accountant, reports that 50 mg of 5-HTP does reduce
his problems with premature ejaculation.

5-HTP and neurological disorders

Any condition that involves the nervous system can be considered a neurological disorder. Such disorders range from the common tension headache to unusual tumors. Some of these conditions, though, are related to disruptions in brain chemistry. Let's take a look at some neurological conditions in which serotonin, and hence 5-HTP, could play a role.

Ataxia

Several neurological disorders lead to ataxia, a condition in which the patient cannot coordinate his or her muscles when attempting voluntary movement. Ataxia is often due to damage or deterioration of the spine or of the cerebellum, a region in the back of the brain involved in neuromuscular coordination and balance.

To determine whether 5-HTP plays a role in improving the condition of patients with ataxia, researchers at the Department of Neurology, Medical University of Lubeck, Germany, gave thirty-nine patients with various diseases 1,000 mg of 5-HTP a day for ten months. Nineteen patients had Friedreich's ataxia, a spinal disorder; thirteen had cerebellar atrophy, in which only the cerebellum is generally involved; and seven had olivopontocerebellar atrophy, in which the cerebellum and other parts of the brain are affected. The results showed that 5-HTP played no significant role in easing cerebellar symptoms (Wessel, 1995).

Researchers from the Ataxia Research Center and Cerebrovascular Unit, Hopital Neurologique, in Lyons, France, did find 5-HTP to be of some benefit in ataxia (Trouillas, 1995). In a double-blind study, eleven patients with Friedreich's ataxia were given, on average, about 300 mg of 5-HTP three times a day and then compared with eight patients who were given a placebo. In the treatment group, there was a significant decrease in kinetic scores, indicating an improvement in coordination. The investigators say, "These results demonstrated that 5-HTP is able to modify the cerebellar symptoms in patients with Friedreich's ataxia. However, the effect is only partial and not clinically major."

An earlier study by Dr. P. Trouillas also supported the use of 5-HTP for cerebellar ataxia (Trouillas, 1982). Twenty-one patients with hereditary ataxia were treated for twelve months with high doses—16 mg per kilogram per day—of 5-HTP. The patients showed significant improvement at the end of the study. Even some patients with multiple sclerosis improved. The researchers think that serotonin is involved in cerebellar neuromuscular control.

Huntington's disease (HD)

This genetically transmitted neurological disorder is characterized by abnormal movements, loss of thinking abilities, and psychiatric symptoms. Aggressive behavior is common, and some researchers suspect that agents which boost serotonin levels could help these patients. In a small study involving two patients with HD, the use of Zoloft (sertraline) was able to reduce their aggressive and irritable behavior (Ranen, 1996). 5-HTP has not been formally tested in HD, but I think such a study would be worthwhile.

Migraine headaches

Classic migraines are throbbing headaches that occur periodically on one side of the head. They are often associated with nausea, vomiting, sensitivity to light, and blurring of vision. These headaches may occur as a consequence of dilation and constriction of certain blood vessels in the neck or head. Attacks can be triggered by emotional or physical stresses, lack of or excess sleep, missed meals, certain foods such as chocolate or alcoholic beverages, and fluctuating hormone levels.

The idea that serotonin may be involved in this condition comes from studies with medicines that help ease migraine headaches. For instance, an oral antimigraine pill called Zomig (zolmitriptan) is known to influence certain serotonin receptors in the brain identified as types 5-HT1B/1D (see Appendix A for more details). Also, serotonin is known to be involved in certain brain nerve pathways that inhibit pain.

5-HTP can potentially play a role in relieving migraine. A Spanish study compared the effectiveness of methysergide, a drug used in migraine treatment, with that of 5-HTP in 124 patients (Titus, 1986). The response rate was equivalent in both groups. Those on 5-HTP had fewer side effects, and the greatest benefit of 5-HTP was in reduction of intensity and duration of a migraine attack. It did not, however, reduce the frequency of migraine attacks.

In another study, Italian researchers gave forty patients with several types of headaches 400 mg of 5-HTP a day for two months in a double-blind, placebo-controlled situation (De Benedittis, 1986). There was a significant improvement when 5-HTP was used in patients who had migraine headaches, although there was little response in those with chronic tension headaches.

The third study using 5-HTP in migraine patients was done in 1991 by Swiss researchers. Thirty-nine patients with migraines participated in a double-blind study, which lasted four months, that evaluated the effect of 5-HTP versus that of a drug called propranolol (Maissen, 1991). Treatment with both medicines resulted in a statistically significant reduction in the frequency of migraine attacks. Both the duration of the attacks and the number of pain medicines used were reduced in the propranolol group more often than they were in the 5-HTP group. The researchers state, "Although propranolol, which is considered a reference for the interval treatment of migraine, is more effective, 5-hydroxytryptophan is a possible alternative for many patients."

Since the research on 5-HTP and migraine is limited, it is difficult to give precise dosage recommendations that would be applicable to everyone. I would recommend working with your health care provider to determine the right dose and frequency of 5-HTP and the appropriate frequency of its use. The continuous use of 5-HTP should not exceed three months until we learn more about this nutrient. If 5-HTP does work, and you do want to use it for longer periods of time, consider going off this nutrient for a month or two while you try other medicines. The herb feverfew might help. You could also use some of the pharmaceutical medicines for a month or two and then come back to using 5-HTP. Or it's possible that a lower dosage of 5-HTP may work in association with feverfew and lower dosages of medicines. This area remains to be explored.

Myoclonus

Myoclonus is defined as a chronic spasm or twitching of a muscle or a group of muscles. These spasms can occur in the eyes, soft palate, facial muscles, or diaphragm, and can often occur at the moment of dropping off to sleep. Although occasional myoclonic jerks can occur in anyone, especially when drifting into sleep, the problem is more common in individuals who are prone to seizure disorders (see the next page). The traditional treatment for myoclonus is the use of benzodiazepines, particularly Klonopin (clonazepam). Dr. Jay Lombard, Clinical Assistant Professor of Neurology at Cornell University in Ithaca, New York, tells me that he has had partial success in using 5-HTP for myoclonus, at a dosage of from 50 to 200 mg a day.

Narcolepsy and cataplexy

Narcolepsy is a sudden urge to sleep during the day that occurs at irregular intervals. Cataplexy is defined as a sudden, temporary attack of extreme muscle weakness. A double-blind, placebo-controlled study was done on eleven patients suffering from these disorders (Autret, 1977). Either 5-HTP or a placebo was administered at the rate of 600 mg a day for four weeks. The daily number of cataplectic and narcoleptic attacks did not vary. However, there was a decrease in the duration of daytime sleep and a significant increase in the duration of nighttime sleep.

Ophtalmoplegia

In this disorder, the motor nerves of the eye become paralyzed. Ophtalmoplegia can be caused by a variety of disorders, including thyroid conditions and strokes. Two patients with a rare form of ophtalmoplegia, called progressive supranuclear palsy (PSP), were treated with several medicines, including 5-HTP (Yukitake, 1996). The administration of 5-HTP at a dose of 600 mg a day improved their horizontal gaze, but failed to improve their other neurological symptoms, such as gait disturbances.

Seizures

A seizure is basically a temporary disturbance of brain function due to an abnormal discharge of nerve activity. About 0.5 percent of the population suffers from seizure disorders, in which seizures occur repeatedly. There are several types of seizure disorders, including partial, complex, generalized, absence (or petit mal), tonic-clonic (or grand mal), and myoclonic. The symptoms produced can range from odd smells and sensations to convulsions and loss of consciousness.

It has been suspected that low levels of serotonin activity may contribute to certain types of seizures, and a small study done in 1970 showed 5-HTP to have some anticonvulsant effect (De La Torre, 1970). To further test this hypothesis, researchers at the Department of Neurology at George Washington University in Washington, D.C., gave 5-HTP to six patients with progressive myoclonus epilepsy, or PME (Pranzatelli, 1995). 5-HTP did not provide any statistically significant improvements in these patients. The researchers say, "Our results suggest a [serotonin] abnormality regardless of the underlying etiology of PME, but one that seldom responds to acute treatment with L-5-HTP."

The role of 5-HTP in the treatment of other seizure disorders has not been evaluated.

Torticollis

Spasmodic torticollis is characterized by the tendency of the neck to twist to one side. This condition usually starts between the ages of twenty-five and fifty. Initially, the neck twisting occurs only occasionally, but it can eventually progress to a permanent state. Although occasional remissions may occur, this disorder can last a lifetime.

A case of very severe spasmodic torticollis is described in the medical literature (Disertori, 1982). The condition was present for eighteen months, and the head of the patient was so twisted that he could only see backwards. The twisting of the neck interfered with the nerves leading from the neck to the arm, which led to problems with sensation and motor control of the arm. Therapy with 5-HTP decreased the patient's symptoms.

The role of 5-HTP in treating childhood disorders

There are several psychiatric and neurological disorders that generally start in childhood. The two most common are attention deficit hyperactivity disorder (ADHD) and autism.

Attention deficit hyperactivity disorder (ADHD)

At least 5 percent of children in kindergarten may have symptoms consistent with ADHD, including an inability to focus on tasks, poor impulse control, and hyperactivity. This condition is more prevalent in boys, and at least a quarter of these children continue to experience symptoms throughout adulthood.

One of the hallmarks of ADHD is sleep disturbance. These children often take an hour to fall asleep, have trouble staying in bed, and frequently wake their parents. Daytime stimulants, such as Ritalin (methylphenidate) and Dexedrine (dextroamphetamine), are often prescribed.

As we saw in Chapter 7, 5-HTP can help induce sleep and could be occasionally used in a dose of from 10 to 25 mg for this purpose. This nutrient can also provide a calming effect. Until long-term studies are done, it would be best to limit the continuous use of 5-HTP to less than three months. However, if used occasionally, such as once or twice a week, 5-HTP can be taken for prolonged periods. See Chapter 10 for interviews with doctors who have used 5-HTP in children with ADHD.

Autism

Several developmental disorders can occur in childhood involving abnormalities in communication, social development, and a restricted repertoire of activities and interests. Autism is one of the most common of these childhood disorders, occurring in one or two children out of a thousand (Rapin, 1998). Additional features can include aggressive or injurious behavior towards self and others, and repetitive thoughts and behaviors. About one-third of these children experience epilepsy by adulthood. Although autism is often considered to be a childhood disease, the majority of patients have significant and incapacitating symptoms throughout their adult lives.

Abnormalities in several brain chemicals have been described in autistic patients, but the most consistently implicated neurotransmitter is serotonin. Manipulation of serotonin levels has given us information about the role of this brain chemical in influencing autistic symptoms. When a group of seventeen adults with autism were given an amino-acid mixture that lacked tryptophan, the precursor to 5-HTP and serotonin, there was a lowering of blood tryptophan levels, along with a worsening of agitation, aggression, and impaired social behaviors (McDougle, 1996a). Moreover, therapy with SSRIs has helped some children by decreasing aggression, improving social relatedness, and reducing repetitive thoughts and behaviors (Cook, 1992; McDougle, 1996b). While no studies have been done on the use of 5-HTP in autism treatment, the apparent link between autism and serotonin means that 5-HTP may be an additional nutrient to be studied.

Summary

As we've seen, there are a number of disorders associated with serotonin. This means that 5-HTP can, potentially, play a role in these situations.

However, limited research makes it difficult to provide guidelines and recommendations on which conditions could be helped by 5-HTP therapy. I hope that additional studies in the next few years will give us some clues, especially on the use of 5-HTP in combination with other natural and pharmaceutical therapies. In the meantime, patients, in consultation with their health care providers, may wish to explore the use of 5-HTP in conditions associated with serotonin disturbances.

CHAPTER 10

WHAT THE EXPERTS
THINK OF 5-HTP

A number of health care providers have used 5-HTP, and tryptophan, in their practices. Here's a sampling of their opinions. (For the opinions of people who have used 5-HTP themselves, see Appendix C.)

Stan Bazilian, M.D., of Philadelphia, Pennsylvania, is a psychiatrist who uses nutritional therapies in his practice, along with pharmaceutical medicines.

> *I have used 5-HTP in a few patients and some of the benefits have been similar to those of the SSRIs. Most patients have felt an improvement in mood.*

Susan Busse, M.D., of Palatine, Illinois, practices nutritional medicine.

> *I don't have much experience with 5-HTP, but I've used tryptophan successfully, especially for insomnia and depression. My dosage for insomnia is 500 mg. Sometimes, I add vitamin B_6, since B_6 is a coenzyme in the conversion of tryptophan to serotonin. Valerian and tryptophan can be a good combination.*

Tryptophan can be used during the day in cases of anxiety. For depression, I use tryptophan at night, and only prescribe additional amounts in individuals who have an agitated or anxious form of depression. Daytime use of tryptophan won't help someone whose depression is associated with fatigue. Instead, I have them take an amino acid in the morning, such as tyrosine or phenylalanine.

I've used tryptophan in children who have attention deficit disorder, at a dosage of 200 mg twice a day, in addition to a nighttime dose. It calms them down. This amino acid can also be useful in Tourette's syndrome [which is marked by uncontrollable gestures, twitching, and foul language]. My dose is 200 mg two or three times a day.

I love tryptophan. It's one of my most prescribed natural nutrients.

Barry Elson, M.D., Medical Director of Northampton Wellness Associates in Northampton, Massachusetts, specializes in nutritional and environmental medicine.

I started prescribing tryptophan back in 1979 for insomnia and anxiety. I mourned the loss when it was pulled off the market in 1989. About three years ago, tryptophan quietly became available as a prescription medicine from compounding pharmacies.

Since 5-HTP became available, I don't need to prescribe tryptophan anymore, except for an occasional patient who does better on tryptophan. In the past three years, I've been recommending 5-HTP, mostly for insomnia, anxiety, and depression, although I occasionally use it for myoclonus, dementia, and fibromyalgia.

Insomnia—I generally use a dosage of 50 mg about a half-hour to an hour before bed. I sometimes use it in combination with calcium, magnesium, hops, passionflower, valerian, and even a small dose of melatonin. I recommend the 5-HTP be taken on an empty stomach and to wait at least twenty minutes before eating. Side effects are rare, but include nausea and vivid dreams. Hangover from 5-HTP is less than that of melatonin. I also have to mention that none of the natural sleep aids are 100 percent consistent in their sleep induction and maintenance.

Anxiety—A 50 mg-dose at bedtime is a good way to start. If need be, I'll recommend my patients take an additional one or two doses during the day, up to a maximum of 200 mg a day. Patients who have anxiety are usually wired and they don't get sleepy from the 5-HTP. I tell my patients to start with small dosages and titrate upwards as needed. I would say 5-HTP is moderately effective for anxiety. Sometimes, I will add valerian or a calcium-and-magnesium combination. More severe cases of anxiety will need a pharmaceutical medicine.

Depression—I'll rarely use 5-HTP by itself for depression. It can be combined with other nutrients, such as St. John's wort or the amino acid tyrosine. 5-HTP works very well in combinations. If I'm using St. John's wort, I'll start with one pill of the standardized 300 mg in the morning and gradually increase the dose to three times a day, if needed. 5-HTP is usually initially given at nighttime but can be additionally taken two to three times during the day, at a dose of 25 to 50 mg. If I'm going to use tyrosine, I'll start with 250 mg in the morning. The idea with combinations is to start low and increase the dose upward gradually. In severe cases of depression, SSRIs are very helpful.

*Obesity—5-HTP could work as an appetite sup-
pressant, but I prefer focusing on lifestyle changes
such as exercise and diet. One nutrient that I find helpful in this situation is hydroxycitrate. 5-HTP would be
helpful in obese patients who also have an anxious
form of depression.*

*I have some patients who have taken 5-HTP regu-
larly for two to three years without any problems.*

*As to children, I use 5-HTP in cases of hyperactivi-
ty, although I find GABA (gamma-amino-butyric acid)
also to be helpful. Older kids who have trouble sleeping
can occasionally use 5-HTP. For obsessive-compulsive
disorder, I'll sometimes use inositol, folate, and DMG
(dimethyl-glycine), and 5-HTP.*

Dale Guyer, M.D., Director of the Center for Innovative
Medicine in Indianapolis, Indiana, focuses on nutritional
therapies and antiaging medicine.

*I've used 5-HTP for the past two years in about two
hundred patients. For fibromyalgia, I start with 50 mg
at night and 25 mg once or twice a day. At least 50 percent are helped. I also use 50 mg at night for patients
with chronic fatigue. In cases of insomnia, 50 mg works
well, although sometimes 75 mg may be needed.*

*In cases of attention deficit disorder, I prescribe
about a third of the adult dose at night. I sometimes
combine the 5-HTP with piracetam. If a child is very
hyperactive, I have them take 10 to 25 mg of 5-HTP
during the day. The 5-HTP can slow down these children for quite a few hours.*

*I haven't yet used 5-HTP in obesity, but it has
helped some patients in their attempts to quit smoking.*

As to individuals with depression, I would say 5-HTP is effective at least 50 percent of the time. I've used it in combination with deprenyl.

Kava can be combined with 5-HTP in cases of daytime anxiety. Low-dose 5-HTP is great for high-strung individuals for daytime use.

Side effects that patients have reported include headaches, stomach upset, vivid dreaming, and nightmares. There have been no allergic reactions and no vomiting.

As a side note, I have found creatine helpful with congestive heart failure. This nutrient can also help some patients with Down's syndrome have better muscle tone.

Peter Hauri, Ph.D., is the Director of the Mayo Clinic Sleep Disorder Laboratory in Rochester, Minnesota.

I'm not concerned that the occasional use of pure 5-HTP for insomnia would have any detrimental physiological consequences.

Joseph P. Horrigan, M.D., board certified in Child Psychiatry, is an Assistant Professor at the University of North Carolina, Department of Psychiatry, Developmental Neuropharmacology Clinic in Chapel Hill.

I generally treat children and teenagers who have moderate to severe cases of autism, attention deficit disorder, seasonal affective disorder, recurrent depression, and Tourette's syndrome. I find 5-HTP partially helpful in these conditions. I prescribe the first dose of 5-HTP in the evening and another dose at bedtime. Kids generally don't get sedated from the 5-HTP during the day as much as the adults.

> *More and more parents are coming to our center armed with knowledge about nutritional therapies that they're reading in consumer magazines and books. The public is getting a lot more sophisticated about alternative ways to treat medical and psychiatric conditions.*

Jeanne Hubbuch, M.D., of Newton, Massachusetts, integrates traditional and alternative medicine in her private practice.

> *Back in the 1980s, before the ban, I was prescribing tryptophan for insomnia and depression. A year or two ago, I started prescribing 5-HTP to my patients and the results have been similar to those obtained with tryptophan.*

> *5-HTP works well for insomnia. Occasionally, some patients feel wired on it and I'm not sure why.*

> *As to depression, 5-HTP is often not effective by itself and I add tyrosine in the morning along with B vitamins. I haven't tried 5-HTP as an appetite suppressant.*

> *In children, 5-HTP helps with ADHD. They become calmer. My dose depends on the child, generally ranging between 25 and 50 mg.*

> *I've had patients on 5-HTP for many months without any problems.*

William Kracht, D.O., FAAFP, of Quakertown, Pennsylvania, is in private practice.

> *I've used tryptophan in about fifty patients. In cases of fibromyalgia, I'll use a combination of St. John's wort, magnesium, and tryptophan. I don't think these nutri-*

ents address the primary cause of the disease, but they do provide partial symptomatic relief. Fibromyalgia is very difficult to treat.

For depression, I've prescribed 5-HTP in combination with tricyclics, St. John's wort, and even tyrosine.

Richard Kunin, M.D., of San Francisco, California, is President of the Society for Orthomolecular Health-Medicine.

I have used tryptophan for insomnia, depression, and obsessive-compulsive disorder with some success. So far, I've only used 5-HTP in a dozen patients. I have recommended it for autistic children at bedtime and in the morning. It helps them sleep and they behave better. I also provide these children with magnesium, B vitamins, especially B_6, and dimethylglycine at 50 mg twice a day.

Jay L. Lombard, D.O., a Diplomate in Neurology of the American Board of Psychiatry and Neurology, is Clinical Assistant Professor of Neurology at Cornell University in Ithaca, New York.

I've had success with 5-HTP at a dose of 50 to 100 mg in the evening in cases of myoclonus. For restless leg syndrome, if 100 mg of 5-HTP is not enough by itself, I'll add 0.5 mg of Klonopin (clonazepam). I've had a soap opera actress who has responded well to this combination. For autism, 25 mg of 5-HTP during the day can sometimes be helpful in making the children calmer.

Larry Pastor, M.D., Clinical Assistant Professor of Psychiatry and Human Behavior at George Washington University Medical Center in Washington, D.C., is an expert in the psychopharmacologic therapy of anxiety and depressive disorders.

> *There are differences between the different SSRIs. Prozac is less sedating than Zoloft, which is less sedating than Paxil. I generally prescribe Prozac in the morning; Zoloft can be taken anytime of the day, while Paxil is best taken at night. Effexor is shorter acting, and it's taken two or three times during the day. Luvox is taken in the evening.*

> *Based on the fact that 5-HTP can be converted in the body into serotonin, I can theoretically speculate that it could have a role to play in obsessive-compulsive disorder, attention deficit disorder, panic disorders, bulimia, anxiety disorders, premature ejaculation, certain sexual disorders, and perhaps premenstrual syndrome. I'm not sure about anorexia, since there are a number of complicating psychosocial and body-image problems associated with this condition.*

> *It's interesting to note that only one or two percent of the neurons in the brain use serotonin. However, this neurotransmitter controls critical functions, such as mood, behavior, and sexual function.*

> *I'm not overly concerned that occasional use of 5-HTP, by its conversion into serotonin, would cause heart valve problems. SSRIs have been used for many years and we haven't found them to affect heart valves, nor do they have much of an influence on the cardiovascular system as do the tricyclic antidepressants. No arrhythmias are noted with SSRIs and changes in blood pressure are rare.*

Christian Renna, D.O., maintains private practices in Dallas, Texas, and Santa Monica, California, and specializes in preventive and nutritional medicine.

5-HTP is potent—a dream come true, one of the most useful over-the-counter nutrients that we have. I use it in cases of anxiety. Since 5-HTP can induce sleepiness if taken in high doses during the day, I start my patients with 5 mg and then gradually increase the dose to a desirable effect. 5-HTP can be mixed with a glass of warm water. It can also be mixed with cinnamon powder and used sublingually [under the tongue], thus bypassing the gastrointestinal system. I've also successfully treated some patients with anxiety by combining it with kava or GABA. I tell my patients who have anxiety that 5-HTP will make the "important things" in their lives that they worry about less important.

As to appetite suppression, I tell my patients who want to lose weight to place 10 mg of 5-HTP under the tongue and take another 20 to 30 mg with warm water a half-hour to an hour before a meal. I think too high a dose of 5-HTP can be counterproductive and cause an imbalance in the norepinephrine and dopamine system, leading to a nonserotonergic urge to eat.

I do believe 5-HTP is effective for weight loss, but it takes time to learn how to use it. The dosage and timing have to be individualized. A patient should be told that the dosage and timing could change with environmental circumstances. A good combination would be to use 5-HTP in a low dose, such as 10 mg, with small doses of ephedrine and caffeine: for instance, 10 mg of 5-HTP with 10 mg of ephedrine and 100 mg of caffeine.

Using 5-HTP as an antidepressant by itself may be limited to those who have an anxious or agitated type of depression.

The side effects are easily predictable and dose-dependent. They include nausea and fatigue.

Douwe Rienstra, M.D., of Port Townsend, Washington, is in private practice.

> I haven't used 5-HTP, but I used to prescribe trypto-
> phan in the 1980s. Since it became available through
> compounding pharmacies about two years ago, I
> restarted prescribing it for depression and insomnia at
> a dose of about 1,000 mg at night. Patients haven't
> noticed any significant side effects. I can sometimes
> avoid using SSRIs in individuals with depression by
> using tryptophan. It sometimes takes two weeks or
> more for the mood-elevating effects to be noticed.

Priscilla Slagle, M.D., is a psychiatrist who maintains private practices in both Encino and Palm Springs, California. She is the author of *The Way Up From Down*.

> I have an enormous amount of experience with tryp-
> tophan and less so with 5-HTP. I find tryptophan is
> more consistent in maintaining sleep. 5-HTP does not
> seem to be as sedating as does tryptophan, and more
> people wake up in the middle of the night when they're
> on 5-HTP.
>
> I have used tryptophan for insomnia for many
> years at a dosage of from 500 to 3,000 mg with no sig-
> nificant side effects. I have one patient with depres-
> sion who's been on 2,000 mg of tryptophan for twen-
> ty-six years without any problems. In 1989, when she
> heard the FDA was going to ban it, she bought herself
> a large supply that carried her through for a while
> and then she got some more from a veterinary supply
> company.
>
> In addition to tryptophan, I occasionally use 1,000
> mg of the amino acid taurine around dinner and anoth-
> er 1,000 mg at bedtime. Magnesium works well, too. I

*find both taurine and magnesium to have antiarrhyth-
mic abilities [preventing heart irregularities].*

*In cases of depression, I recommend tyrosine, a B-
complex pill, 20 mg of pyridoxal-5-phosphate [a form of
vitamin B_6], magnesium, and tryptophan. Even though
phenylalanine converts into tyrosine, I don't use it
much since it can also convert into phenylethylamine.
This brain chemical is very stimulating, and can cause
agitation, increase blood pressure, and increase the possibility of arrhythmias. I'll also add small amounts of
DHEA or pregnenolone in patients middle-aged and
older, if testing shows they are deficient.*

Karlis Ullis, M.D., is Clinical Professor at the University of
California in Los Angeles, and maintains a private practice
in Santa Monica, California. Dr. Ullis specializes in nutritional therapies, sexual medicine, and antiaging medicine.

*I have found 5-HTP and tryptophan to be helpful in
patients who are impulsive, and lack control in terms
of overeating, gambling, and other compulsions. 5-
HTP appears to have a tempering effect on their behavior. Some patients have a better response to tryptophan,
while others respond better to 5-HTP.*

Nathan D. Zasler, M.D., is editor-in-chief of *Neuro-
Rehabilitation: An Interdisciplinary Journal*, and is medical
director of the Concussion Care Centre of Virginia, near
Richmond.

*I treat patients with concussions and acquired brain
injury. I've used 5-HTP in about forty patients with
post-concussion syndrome who have difficulty sleeping, or who have irritability or emotional lability
[instability]. 5-HTP helps with sleep initiation. The*

usual dose is about 50 mg. I would say about half of the patients have a positive clinical response. Melatonin is also helpful as a sleep aid.

Serotonergic precursors should certainly be considered as monotherapy or as adjunctive therapy with other medicines in persons with acquired brain injury. Theoretically, many symptoms that arise after a concussion appear to be due to serotonin insufficiency. It makes sense to use serotonin agonists like 5-HTP. I think 5-HTP can partially substitute for sedating medicines such as Desyrel (trazodone, a sedating antidepressant). I've sometimes combined low doses of both 5-HTP and trazodone with good response.

I haven't had too much luck using 5-HTP alone for myoclonus. I use more traditional drugs like clonazepam, mysoline, and divalproex, but I plan to try the combination. It's simplistic to expect that complicated neurological disorders will respond to one magic bullet. We would have a better effect combining more than one medicine or nutrient in treating impairments related to certain neurologic and/or neuropsychiatric disorders.

Summary

Clinical experience with 5-HTP is limited. Therefore, as yet, there is no consensus on its ideal use, by itself or in combination. But in general, most physicians have noted that 5-HTP plays a positive role in disorders related to mood, anxiety, appetite, and sleep. In the next chapter, I'll tell you how to use 5-HTP safely and effectively.

CHAPTER 11

USING 5-HTP—
THE RIGHT DOSE AND
THE RIGHT TIME

You can't be an expert chef just by buying a few kitchen utensils and a cookbook. It takes time to learn how to combine different ingredients in the right proportions, what condiments to add, and what cooking temperature to use.

This is also true when taking supplements. You can't be an expert in natural therapy after reading a couple of articles in magazines, buying a few bottles of supplements, and swallowing the pills. It takes a great deal of knowledge, experimentation, and time to become an expert. Even though I've studied nutrition for twenty years, have treated thousands of patients, and taken, at one time or another, almost every natural supplement on the market, I'm still accumulating additional knowledge every day.

Furthermore, you must take into account that your personal biochemistry could very well be different than the norm. In fact, no two individuals are exactly alike in their response to supplements and medicines. There could well be differences in gastrointestinal absorption, liver metabolism, transport through the blood-brain barrier, and a host of other important factors that would influence your response. Therefore, when you start taking 5-HTP, please realize that the guidelines provided in this book may or may not apply to you. If you are sensitive to medicines, always start with the lowest possible dosages. If you normally hardly feel the effects of pills that you take, I would still recommend that you start low and then gradually increase the dosage as you become more familiar with the effects.

In this chapter, I will provide guidelines on how you can best use 5-HTP, including how you can combine it with other supplements and medicines to treat insomnia, anxiety, depression, and obesity. A health care provider familiar with natural therapies should be your guide while you are taking these supplements. But first, let's discuss a few practical aspects of taking 5-HTP.

How is 5-HTP available?

You can purchase 5-HTP over the counter in vitamin stores, pharmacies, retail outlets, and mail order firms. It is sold in a variety of dosages, but most commonly in 50- and 100-mg capsules, although I hope some vitamin companies will also market 10- and 25-mg capsules. Read the label on the bottle carefully. It may say "5-HTP" on the front, but if you read the back of the label more closely, you may find that other nutrients and herbs have been added. If your main intention in taking 5-HTP is for insomnia relief, you don't want to take a product with added stimulants. If your main intention is to use 5-HTP during the day as an appetite suppressant, you don't want a product with added sedative herbs or nutrients.

As a rule, content accuracy for over-the-counter products is very good. Our research center tested a number of DHEA products in 1996, and found all to contain 87 to 101 percent of the amount of DHEA claimed on the label. (The results are posted on the Internet at www.raysahelian.com.) The National Nutritional Foods Association, in Newport Beach, California, regularly tests over-the-counter products sold in vitamin stores. They have found that, in most cases, companies are following good practices. One concern that I have about 5-HTP is that the raw material is expensive, and if suddenly there's a large demand for it, a shortage could develop. This could lead a small number of companies to place less of it in each capsule until the shortage is corrected.

Is 5-HTP best taken with food or on an empty stomach?

I have personally experimented with 5-HTP and have found that the effects are more noticeable and consistent when 5-HTP is taken without food. If I take 5-HTP an hour before bedtime on an empty stomach, I notice a yawn within fifteen minutes or less. I suspect 5-HTP is absorbed much more quickly when there is nothing else in the stomach, and when no amino acids are competing with it to cross the blood-brain barrier (see Chapter 3).

However, some studies have shown 5-HTP to be clinically effective when taken with a meal. A 1991 study by W. Poldinger and colleagues found 100 mg of 5-HTP taken three times a day with food to be an effective antidepressant. Taking 5-HTP with meals could reduce the gastrointestinal side effects, but a higher dosage may be required to provide a noticeable effect.

Is it healthier to have quick absorption, and a subsequent fast rise in blood and brain 5-HTP levels, or slower absorption resulting from the presence of food in the stomach? The answer is not known. My opinion as of now is that taking a smaller amount of 5-HTP on an empty stomach could be equivalent to taking a higher dose on a full stomach. For maximum absorption, take 5-HTP at least two hours after a meal and at least a half-hour before you eat again. However, the best way to find out which works better for you is to try 5-HTP with and without meals. Another option is to take the 5-HTP ten to twenty minutes before eating. With this approach, some of the 5-HTP would already have been absorbed, while the rest would be absorbed more slowly.

Will 5-HTP make me sleepy or alert?

Sometimes it's difficult to predict whether 5-HTP will cause sedation or alertness. A number of factors are involved, including dosage, recent meal content, the time of day you take the 5-HTP, the supplements or medicines you are currently on, your age, and perhaps your menstrual cycle if you're a woman.

Generally, when 5-HTP is taken at night, half an hour or an hour before bedtime, it causes sedation. However, when it's taken during the day, it can cause either sedation or alertness. Generally, a dose of 25 mg or less will not cause sedation, but it's possible that a higher dose could make you sleepy. Whether 5-HTP makes you sleepy or alert will depend on your body chemistry. If you're the type of person who is very wired or alert, you will feel relaxed on 5-HTP, but not sleepy. If you're already sluggish, you may have the urge to sleep. Another factor to consider is that 5-HTP can work in cycles. You may feel alert for a few hours and then sleepy, feel alert again, and so on.

The type of meal you consume could also make a difference. If you take 5-HTP before a carbohydrate meal, you may feel very sleepy, while if you take it during the day along with small, frequent meals consisting of mostly protein, you may find yourself alert. Each of us has cycles throughout the day where we feel more alert or more sedated. The effect from 5-HTP could depend on the timing during this cycle. As a rule, taking 5-HTP in the morning, when one tends to be most alert, will not make a person sleepy, but taking it during a midafternoon or early evening slump could induce the urge to take a nap.

If you take any type of stimulant during the day, even if it's a cup of coffee, this can partially prevent or reverse 5-HTP's sedative effect. You may also find a smaller amount of 5-HTP, such as from 10 to 25 mg two or three times a day, to provide more consistent effects than a large amount, such as 100 mg, taken at one time. Again, these are individual variations.

Using 5-HTP for appetite control

5-HTP does help many people in their attempts to consume fewer calories. Studies published on 5-HTP in weight reduction have shown this supplement to be effective (see Chapter 4). However, the dosages used in these studies were pretty high. I feel uncomfortable using such high dosages, and would prefer to see you start low. Follow these recommendations while under medical supervision:

- If you are trying to lose weight and also have difficulty sleeping, start with from 25 to 50 mg of 5-HTP about an hour before bedtime for a few nights. Hopefully, the nightly use can have a residual effect the next day in terms of appetite suppression. After a few nights of taking 5-HTP, I would recommend you take this nutrient at night no more than twice a week.

- Next, add an additional 5-HTP dosage of from 10 to 25 mg during the day, about an hour before a meal.

- If the premeal dosage is not enough, you could increase it to between 25 and 50 mg once or twice a day on an empty stomach. One problem with this higher amount is the possibility of inducing daytime sedation. The sleep-inducing aspect of 5-HTP would limit its usefulness unless it were combined with a medicine or nutrient that causes alertness.

• You could consider drinking a cup or two of coffee in the morning. Don't drink coffee later than midday, since caffeine can last in the system for many hours and late afternoon or evening coffee could interfere with sleep. Caffeine could have a synergistic appetite-suppressing effect with the 5-HTP.

• If you're not a coffee drinker, try either black or green tea. Other options include taking a guarana pill, which contains caffeine.

• If you don't have any heart problems or high blood pressure, a small dose of ephedrine, such as 4 mg once or twice a day, could help you with appetite suppression. Avoid any daily amount greater than 10 mg. You must be supervised by a health care provider. Ephedrine has side effects that include heart irregularities.

• Another option, especially if you have a mild to moderate case of depression, is to take a 300-mg St. John's wort pill in the morning, two pills a day maximum.

• Additional natural supplements to try include the amino acids phenylalanine or tyrosine. For either, the starting dosage would be from 100 to 250 mg, taken in the morning. High dosages can lead to anxiety, restlessness, and palpatations. Do not combine ephedrine with these amino acids.

• You should add fiber to your diet. One option is to stir a teaspoon of psyllium powder or seeds into a glass of water, and take this mixture at the beginning of each meal. For more details, see Chapter 4.

The maximum dosage of 5-HTP I recommend for weight-loss purposes is 50 mg three times a day. I would also recommend not taking 5-HTP for a period longer than three months without taking a break. I even prefer you take a break for a week after a month's use. In addition, skip taking 5-HTP two days a week. We just don't have enough studies to tell us whether this serotonin precursor is safe for prolonged use.

The idea of using 5-HTP is to provide you with enough appetite control to get you started on a weight-loss program. The ultimate goal, though, is for you to eventually begin an exercise regimen, and to find behavior-modification methods that allow you to decrease your food consumption. Also, be sure to consume adequate amounts of protein, so as to avoid losing muscle mass, and be careful about adding high dosages of stimulants, since they can cause heart palpitations.

One temptation that some patients face is to try to lose the weight too quickly by taking high dosages of 5-HTP. It is much easier on the body if you attempt to lose weight slowly, thus placing less stress on your organs and tissues.

There are risks anytime one uses natural nutrients or combinations for weight-loss purposes. However, the risks inherent with the use of natural substances are generally less than the risks associated with pharmaceutical medicines currently prescribed for appetite suppression. We need to also consider the fact that untreated obesity can have detrimental effects on the body. Neither the physicians interviewed in this book nor I have much experience in combining 5-HTP with the currently available weight-loss drugs.

The multisupplement approach to fighting depression

Natural supplements are best suited for mild to moderate cases of depression. There are quite a number of options at our disposal, and there are no standards that have been established for natural depression treatment. If you go to a dozen doctors who use natural therapies, you are likely to encounter just as many different ways to treat your depression.

The following is the approach that I have developed in treating patients with depression. (For more information, see Chapter 5.) Your health care provider may wish to follow this guideline, adapt it to your specific case, or decide that your particular case requires a completely different approach:

• Start with one 300-mg pill of St. John's wort with breakfast, using the standardized 0.3 percent hypericin extract. You can find these standardized extracts practically everywhere St. John's wort is sold. (See *St. John's Wort: Nature's Feel-Good Herb* in Appendix B.) Also, take a B-complex vitamin that contains about two to five times the Recommended Daily Allowance (RDA).

• If, after three days, you don't notice enough mood elevation, you now have two options. If your depression is associated with anxiety, add 5-HTP at a dosage of from 25 to 50 mg before breakfast, lunch, or dinner. If your depression is associated with lethargy or tiredness, add a second dose of St. John's wort with either breakfast or lunch. One of the side effects of St. John's wort is over-alertness, which can lead to insomnia if the stimulation persists late into the night. 5-HTP may counteract this alertness.

• If you find that your anxiety persists, you can add a second dose of from 25 to 50 mg of 5-HTP during the day. One of the disadvantages of taking 5-HTP during the day is that it can induce sleepiness in higher dosages. Therefore, St. John's wort is a good option to use with the 5-HTP, in that it partially, or mostly, prevents this sleepiness.

The maximum daily dose of 5-HTP that I recommend for depression at this time is 100 mg. Also, until more studies are available, I would limit the continuous use of 5-HTP to no more than three months. After a break of two to four weeks, your health care provider can restart the 5-HTP if needed. Just to be on the safe side, skip taking 5-HTP one or two days a week.

So far, you're on B-complex vitamins, St. John's wort, and 5-HTP. If, after a month on these supplements you need additional mood elevation, you can add one of the following:

- The amino acid tyrosine can be converted into dopamine and norepinephrine, thus increasing energy and alertness. Start with a low dose of 100 mg in the morning.

- Coenzyme Q_{10} can be used at a dose of 10 to 30 mg early in the day to increase energy.

- Kava is an excellent herb to take if your depression is associated with anxiety. The dosage would be between 70 and 100 mg of the kavalactones. Kava can be taken any time of day, but late afternoon or early evening are good options. Some people will feel too alert if a high dose is taken at night.

- Ginkgo biloba, at 40 mg a day in capsule form, can be helpful when taken before noon. It is also available as a tea.

- If you're over the age of forty-five, you can temporarily use either DHEA or pregnenolone at a dosage of from 2 to 5 mg in the morning. Only use these hormones five days a week, and take one week off a month. Alternately, you can take these hormones every other day. I prefer hormones to not be used continuously without breaks. Preg and DHEA can enhance sexual drive and enjoyment, and can provide a sense of well-being. Preg can enhance visual and auditory perception. High dosages can, in some people, cause acne, headaches, irritability, heart palpitations, and accelerated hair loss. (See *DHEA: A Practical Guide* and *Pregnenolone: Nature's Feel Good Hormone* in Appendix B.)

Additional supplements and herbs that can increase energy or mood include carnitine, acetyl-L-carnitine, ginseng, S-adenosyl-methionine, and phosphotidylserine. Whenever you add any supplement, always use low dosages in order to avoid overstimulation. If you have a propensity for heart irregularities, be careful about adding too many stimulants, because some of them can trigger palpitations.

Using 5-HTP to reduce anxiety

Several nutritional and herbal options are available to treat mild to moderate cases of anxiety. It may take time to find the ideal supplement combination that works for you, but, if you persist, you will find nutrients and herbs that work best for your particular situation. (For more information, see Chapter 6.) The following are some guidelines you may wish to follow while under the care of your health care provider:

- Start with 25 mg of 5-HTP a day. The timing will depend on whether you feel most anxious in the morning, afternoon, or evening.

- If 25 mg once a day is not enough, you can increase the dosage to 50 mg. There may be certain days where you are very anxious. It is acceptable on these days to take higher amounts, such as 100 to 150 mg.

- If you need additional help, add kava to your regimen. The dosage could range from 70 to 100 mg of the kavalactones once, twice, or three times a day. Kava can induce simultaneous mental alertness and muscle relaxation. (For more information, see *Kava: The Miracle Antianxiety Herb* in Appendix B.)

- If you still need help, add 200 mg of magnesium twice a day.

- The amino acid taurine can help in inducing relaxation. Take 500 mg on an empty stomach.

- Additional options include drinking teas during the day made with valerian, hops, passionflower, or chamomile, either alone or in combination.

- If all else fails, the occasional use of a pharmaceutical antianxiety agent can be justified.

I would not recommend the continuous use of 5-HTP for longer than three months until we learn more about the full long-term effects of this serotonin precursor. Just to be on the safe side, skip your 5-HTP one or two days a week. Whenever you are already on one or more nutrients or herbs, and you plan to add additional ones, do so in a gradual way. Individuals predisposed to panic would do well to avoid stimulants that could precipitate attacks, including yohimbine (Albus, 1992), caffeine (Uhde, 1984; Bourin, 1998), and ephedra, and the stimulating amino acids phenylalanine and tyrosine.

The polytherapy approach to natural sleep

Over many years of practicing medicine, I have gradually come to the realization that the brain can build a tolerance to just about every sleep-inducing agent, and therefore the best way to treat chronic insomnia is to alternate different supplements and medicines. Thus, I see the role of 5-HTP in the therapy of insomnia as a supplement that can provide tremendous benefits if it's used only once or twice a week. (For more information, see Chapter 7.)

If you take 5-HTP at night, especially on an empty stomach, you will most likely feel the onset of a yawn and the urge to go to bed within an hour. The following are my recommendations on treating insomnia with natural supplements. After determining your 5-HTP dosage, use the other supplements on alternate nights:

- Take 25 mg of 5-HTP if you have a mild case of insomnia, or 50 mg if you have a moderate case. This should be taken on an empty stomach about thirty to sixty minutes before bedtime. After a minimum of twenty minutes, you can eat a small to moderate meal consisting of carbohydrates (see Chapter 7). You can take a dose as high as 100 mg if needed, although this would increase the likelihood of having a vivid dream, even a nightmare. Some patients do better taking 5-HTP in the early evening, a few hours before bedtime. Experiment to see which option works better for you.

- Valerian root capsules, in a dosage of between 200 and 500 mg of the concentrated root extract, or valerian tea can be taken about one to three hours before bedtime.

- Hops capsules or hops tea can be taken about an hour or two before bedtime.

- You can drink a combination herbal tea consisting of several herbs, such as chamomile, passionflower, hops, and valerian. Another option is to open capsules of the different herbs, pour them in a cup, and add hot water.

- You can use melatonin once or twice a week in a dose of from 0.3 to 1 mg. You can use any form, including time-release capsules, pills, sublingual lozenges and liquid, or tea. (See *Melatonin: Nature's Sleeping Pill* in Appendix B.)

- Kava, at a dosage of between 40 and 70 mg of the kavalactones, can help some individuals relax and sleep better, although others may find that the kava causes alertness rather than sedation. You could combine the kava with other herbs, 5-HTP, or melatonin. Another option is to take the kava in late afternoon or early evening.

- For moderate or severe cases of insomnia, taking a pharmaceutical sleeping pill once in a while can be justified.

You could certainly experiment with combining a small dose of 5-HTP, from 10 to 25 mg, with one or more of the herbs. Since each of us has a unique biochemistry, it may take trial and error until you find the formula that works best for you. At least once or twice a week, don't take anything, even if it means not getting a good night's rest. Your brain will respond better on subsequent nights to the sleep aids.

If you wake up in the middle of the night and want a fast-acting supplement, open a capsule of 5-HTP, pour the contents into a small dish, and add cinnamon for flavor (a suggestion from Dr. Christian Renna). Then, with a wet finger, you can place this mixture under your tongue for a quick sleep-inducing effect. Don't go back to bed right away, but stay up for at least twenty minutes in a dark room and do something soothing, such as listening to calming music.

Should I take a drug to prevent the conversion of 5-HTP to serotonin in the blood?

Most of the studies involving the use of 5-HTP in treating

depression have combined this serotonin precursor with medicines that block the conversion of 5-HTP to serotonin outside of the brain. These medicines, which include carbidopa and benserazide, are known as peripheral decarboxylase inhibitors (PDIs). As you may recall from the diagram in Chapter 3, 5-HTP can be converted into serotonin by an enzyme called AADC. PDIs block the activity of AADC outside of the central nervous system. This is supposed to allow more 5-HTP to enter the brain and provide the positive central nervous system effects we are seeking. The prevention of the conversion of 5-HTP to serotonin outside of the brain is also thought to decrease, or perhaps avoid, some of the possible side effects of serotonin excess, such as those that occur in the medical condition known as the carcinoid syndrome (see Chapter 12).

Theoretically, it makes sense to use a PDI along with 5-HTP. But practically, are PDIs really necessary? To answer this question, Dr. K. Zmilacher and colleagues, at the Psychiatric University Hospital in Basel, Switzerland, gave thirteen depressed patients 100 mg of 5-HTP, gradually raising this dosage in a few of the patients to 300 mg (Zmilacher, 1988). In thirteen other patients, the 5-HTP was given in conjunction with benserazide. This was an open study, and the researchers continued giving some of the patients their routine antidepressant medications.

Both groups saw similar benefits in terms of mood elevation, which generally occurred within five days. The side effects of nausea and diarrhea depended on the dosage—the higher the dosage of 5-HTP, the more frequent the side effects—and occurred more frequently in patients treated with 5-HTP alone. Anxiety occurred more frequently in those individuals who were on combination therapy. Interestingly, the therapeutic efficacy was not dose-dependent: Many patients did better on the lower dosages.

The researchers conclude,

> *A review of the literature on this subject revealed that 5-HTP given alone was more effective (249 out of 389 patients, 64%) than the combination of 5-HTP with a peripheral decarboxylase inhibitor (93 out of 176 pa-tients, 53%).*

What about the risks of excess serotonin in the blood and tissues? Can PDIs reduce this chance? In one study, when a single oral dose of 5-HTP was given by itself and compared with a dose given in combination with a PDI, there was a significantly higher rise in 5-HTP blood levels, and a decrease in serotonin levels, in those who received the PDI (Magnussen, 1981). However, another study evaluated blood serotonin and 5-HTP levels after continuous treatment with 5-HTP, carbidopa, and benserazide for several weeks. There was an accumulation of serotonin in blood platelets, despite the administration of PDIs (Magnussen, 1982).

At this point, having reviewed all the studies published on 5-HTP, I cannot come to a definitive conclusion as to whether it is preferable in the long run to give 5-HTP by itself or with a PDI. However, when used for a period of three months or less, 5-HTP alone should not present any untoward health risks. Nevertheless, it is up to your health care provider to decide whether the addition of a PDI, such as carbidopa, is necessary in your situation.

To B_6 or not to B_6

Many 5-HTP supplements include vitamin B_6—or pyridoxal phosphate, one of the forms of the vitamin active in the body—in dosages ranging from 2 to 20 mg. In the past, vitamin B_6 was frequently added to tryptophan pills. This vitamin is an assistant, or cofactor, in the eventual conversion of tryptophan to melatonin. It's difficult to say whether the addition of this vitamin to 5-HTP supplements is necessary and beneficial, or unnecessary and potentially counterproductive.

The potential benefits from vitamin B_6 include the fact that it could help 5-HTP be converted more efficiently in the brain to serotonin. However, at the same time, this vitamin could help convert 5-HTP to serotonin in the bloodstream or tissues, which is something we want to minimize.

Another factor to consider is that vitamin B_6 helps in the manufacture of both norepinephrine and dopamine, which could theoretically lead to more of these alertness-causing brain chemicals to be formed. This could be desirable in the daytime, but counterproductive if the 5-HTP is taken at night, since it may interfere with sleep.

These ideas are mostly speculative, since we don't have any studies evaluating the effects of 5-HTP with or without vitamin B_6. You may wish to try 5-HTP both ways to see which option works better for you. We do know that 5-HTP works well by itself. However, I think a B-complex supplement would certainly be a worthwhile addition when taken in the morning with breakfast. You could then take the 5-HTP a few hours later, or in the evening.

Summary

5-HTP is a natural supplement that can be used in the treatment of a number of conditions associated with low brain serotonin levels. 5-HTP's brightest promise lies in finding ways to intelligently combine it with other natural supplements. At this time, until we learn more, limit your daily use of 5-HTP to no more than 100 mg, and limit your continuous use to no longer than three months. Skip your 5-HTP dose a day or two each week.

CHAPTER 12

CAUTIONS, SIDE EFFECTS, AND INTERACTIONS WITH OTHER THERAPIES

The long-term effects of 5-HTP on the human body are not known. There is reason to be cautious, since this supplement plays a role in a variety of bodily tissues.

Cautions and side effects are important issues we must address. There are hardly any psychoactive medicines or natural supplements that are free of side effects. Many consumers have the unreasonable expectation that a supplement should only provide positive effects. This view is unrealistic. If a supplement is powerful enough to cause a noticeable change in mood or behavior, it is also strong enough to cause an unpleasant reaction if misused.

In addition, we must address the issue of interactions between 5-HTP and a number of other substances. In the last few years, dozens of supplements that were gathering dust on the shelves of health food stores have suddenly become hot topics. Countless Americans have started taking herbs, hormones, and the like. You might be one of them. You may wonder, "How does 5-HTP interact with the pills I am already taking?"

In this chapter, I will address these issues. I will also discuss the medical tests you should undergo if you are planning to use 5-HTP regularly for long periods of time.

Cautions in using 5-HTP

First of all, you should be aware that a high daytime dosage of 5-HTP can have a sedative effect in some individuals. This may interfere with driving or operating heavy machinery.

One of the concerns with regular, long-term therapy with 5-HTP is the possibility of tolerance. William Byerley, M.D., from the Department of Psychiatry, University of Utah in Salt Lake City, says, "Another issue is whether 5-HTP increases or decreases serotonergic neurotransmission [serotonin activity]. Short-term administration presumably enhances serotonin activity, but long-term use may cause the opposite effect. Thus, it is possible that the effectiveness of 5-HTP would lessen over time." (Byerely, 1987)

Over the past few months, I have come across a number of other questions and concerns raised by consumers and physicians regarding the safety of 5-HTP. I'll address these concerns and provide some evidence to support or dismiss their validity.

The heart valve concern

The weight-loss medicines known as Redux and Phen-Fen were blamed for causing heart valve abnormalities in some users. It was presumed that the abnormalities—which involved a thickening of the valves—resulted from an elevation in serotonin levels, although there is still debate in the scientific community as to the real cause of this problem. Serotonin itself may not be the culprit, since selective serotonin reuptake inhibitors (SSRIs), such as Prozac, which also increase serotonin levels, are not known to cause such abnormalities.

A link has been made between excessively high serotonin levels for prolonged periods and heart valve problems because of a medical condition known as malignant carcinoid syndrome. Certain cells in the lining of the gastrointestinal system, known as argentaffin cells, are known to create serotonin, among other active compounds. A malignant tumor of these cells can lead to the overproduction of serotonin, and blood levels of serotonin can become extremely high. The symptoms of carcinoid tumor include facial flushing, abdominal cramps and diarrhea, spasms of the bronchial tubes—and problems with the tricuspid heart valve.

Do blood levels of serotonin increase after taking 5-HTP? In a study conducted at the Clinical Pharmacology Unit, Department of Medicine, Royal Infirmary in Edinburgh, Scotland, Dr. Li Kam Wa and colleagues wanted to see if the administration of 5-HTP led to increases in blood serotonin levels (Wa, 1995). Six healthy, young male volunteers were studied in a randomized, crossover trial. The volunteers received 10 mcg per kilogram of body weight per minute of intravenous 5-HTP in a one-hour period, for a total dose of about 40 mg. Blood and urine samples were collected every thirty minutes before, during, and several hours after the infusion.

Interestingly, the 5-HTP administration did not lead to any significant increase in blood serotonin levels. However, there was a significant increase in the amount of serotonin excreted in the urine. This can be explained by the fact that 5-HTP can be converted into serotonin in the kidneys. The researchers state, "The present study confirms our previous observations that 5-HTP markedly increases urinary serotonin excretion. In addition, we have now shown that this occurs without concomitant changes in [blood] circulating serotonin levels. These findings support our hypothesis that urine serotonin, after infusion with a serotonin precursor, is largely derived from intrarenal [kidney] synthesis of serotonin.

"5-HTP is decarboxylated to serotonin by aromatic L-amino acid decarboxylase. This enzyme has a wide distribution with high activity in the kidney and liver. The absence of an increase in circulating serotonin after administration of 5-HTP suggests that serotonin, if produced extrarenally [outside the kidneys] is rapidly metabolized and cleared from the circulation."

However, when one ingests a 5-HTP pill, it can be converted into serotonin in the intestines, and blood serotonin levels could consequently rise. A Japanese study showed that serotonin levels rose when volunteers were given oral 5-HTP at the rate of 3 mg per kilogram of body weight (Kaneko, 1979).

In Chapter 9, I discussed a study by researchers at the Medical University of Lubeck, in Germany, who gave 1,000 mg of 5-HTP to patients with ataxia for a period of ten months (Wessel, 1995). They performed eight electrocardiograms (EKGs) during this ten-month period, and no abnormalities were found. What would be needed is a study in which echocardiographs are performed before and after a few months' worth of 5-HTP supplementation. Unlike an EKG, which records the heart's electrical activity, an echocardiograph is a sonogram that allows doctors to evaluate the size of the heart, and the size and shape of the valves.

At this time, we don't know for sure whether taking 5-HTP continuously for a prolonged period will cause heart valve problems. For this reason, until we learn more, I don't recommend that anyone take 5-HTP daily for longer than three months without taking breaks. If your medical or psychiatric condition requires that you take 5-HTP continuously for months and years, I would recommend having an echocardiograph done every few months, just to be on the extra-cautious side.

Serotonin syndrome (SS)

Serotonin levels that are too high can lead to a condition known as serotonin syndrome (SS). This can occur with the improper use of SSRIs or monoamine oxidase inhibitors (MAOIs). The symptoms include restlessness, confusion, sweating, diarrhea, excessive salivation, high blood pressure, increased body temperature, rapid heart rate, tremors, and even seizures (Martin, 1996; Kirk, 1997). Although SS may lead to death in extreme cases, most patients recover once their medications are stopped.

In addition to many antidepressants, there are additional drugs, both legal and illegal, that could precipitate SS, either by themselves or in combination with other medicines. These include LSD, MDMA ("Ecstasy"), lithium, L-dopa, and buspirone (Martin, 1996). Pseudoephedrine and phenylpropanolamine can also cause problems. The herb ma huang contains ephedrine, which is related to pseudoephedrine, as do some over-the-counter cold medicines.

Combining an MAOI with 5-HTP should be done under close medical supervision. In one study, the combined use of MAOIs with 200 mg of 5-HTP resulted in high blood pressure and emotional instability in one patient (Nicolodi, 1996). However, this study, which went on for one year, did not indicate any other problems to have developed.

Tryptophan can produce central nervous system effects similar to that of mild SS when given in combination with high doses of SSRIs. Symptoms occurred when Prozac, at a dosage of from 50 to 100 mg a day, was taken in combination with from 2 to 4 grams of tryptophan (Steiner, 1986). Keep in mind that most patients take Prozac in dosages of 20 mg a day or less.

SS-like adverse reactions can also occur from excessive doses of 5-HTP. When eleven patients with chronic schizophrenia were given from 2,500 to 9,000 mg a day, they experienced sweating, low blood pressure, increases in motor activity (restless movements), and a temporary increase in psychosis (Wyatt, 1973). These were in addition to the digestive problems that often arise when high doses of 5-HTP are used.

SS has not been reported in the medical literature as a consequence of moderate 5-HTP use. However, we should keep this in mind as a possibility, especially if this nutrient is combined with other stimulants, either natural or prescription. High dosages of St. John's wort may also cause SS. Therefore, when combined, 5-HTP and St. John's wort should be taken in small amounts.

Scleroderma

Also known as progressive systemic sclerosis, scleroderma is a chronic disorder characterized by thickening and hardening of tissues in the body, particularly the skin and internal organs. For instance, the tissues of the esophagus can become hardened, leading to difficulty in swallowing. There can be heart, kidney, and lung involvement.

It is possible that one problem in some patients who are prone to scleroderma is their inability to easily transform serotonin to its metabolite, or byproduct, called 5-hydroxy-3-indole acetic acid (5-HIAA). When tryptophan is given to patients with this condition, their serotonin levels rise, but their 5-HIAA levels do not (Stachow, 1977). This study raises the concern that for patients with scleroderma, or those prone to this condition, it may be best to avoid both tryptophan and 5-HTP.

Another problem could be abnormalities in the metabolism of tryptophan and 5-HTP byproducts. For example, a 70-year-old patient who was being treated for myoclonus developed scleroderma-like symptoms after taking 1,400 mg of 5-HTP and 150 mg of carbidopa for twenty months (Sternberg, 1980). This patient had high levels of a chemical called kynurenine in his blood, suggesting an abnormal metabolism of tryptophan and 5-HTP. After discontinuing the 5-HTP, his symptoms gradually improved. The researchers say, "Our data and studies in the literature suggest that two factors may be important in the pathogenesis of some scleroderma-like illness: high plasma serotonin and the abnormality associated with elevated kynurenine."

The recommendations in this book are that you limit your regular, continuous supplementation to less than 100 mg a day, and to a period of three months. With this conservative approach, any potential problems would be minimized.

Pregnancy and 5-HTP

The four most widely used antidepressants—Prozac, Paxil, Zoloft, and Luvox—do not appear to cause birth defects, according to a study published in the February 1998 issue of the *Journal of the American Medical Association*. We don't know if this means that 5-HTP is safe for pregnant women, since no studies have evaluated the role of 5-HTP in pregnancy. If you are pregnant, your health care provider needs to evaluate the benefits of 5-HTP versus the possible risks.

Other concerns

The elderly metabolize medicines much more slowly than younger people, and thus should take smaller dosages of 5-HTP, at least to start. This is especially true of those older people who have kidney or liver disease; they need smaller amounts taken less often. Anecdotal information indicates that the elderly may be more susceptible to the sedative effects of 5-HTP than the young.

We don't know how 5-HTP influences the thyroid gland, the pancreas, or other glands, although high doses of 5-HTP are known to stimulate the production of prolactin from the pituitary gland. Therefore, anyone with excess prolactin secretion should stay away from 5-HTP at this time.

When rats are intravenously injected with very high doses of 5-HTP (60 mg/kg), less protein is formed in their brains (Lepetit, 1991). The amount the rats received would be equivalent to a 70-kilogram, or 154-pound, human receiving 4,200 mg a day. We don't know how lower dosages of 5-HTP taken regularly might affect protein formation in the brain.

Possible side effects of 5-HTP

Sometimes we don't find out about all the possible side effects of a medicine or supplement until a large number of people start using it. Almost all of the studies done with 5-HTP have used a small number of subjects. As the popularity of 5-HTP grows, we can expect tens of thousands, even millions, of Americans to begin supplementing with it. As a result, we may discover that some individuals will experience side effects of 5-HTP, or of its use with other medicines and supplements, that were formerly unknown.

One relatively long-term study has not shown 5-HTP to pose any major risks. Researchers at the Medical University of Lubeck in Germany gave thirty-nine patients 1,000 mg of 5-HTP a day for ten months (Wessel, 1995). They note,

> Except for minor gastrointestinal side effects in eight patients that did not cause serious problems in the treatment, no relevant side effects were noted with long-term administration of the 5-HTP in a dose of 1,000 mg/day. In particular, none of our patients showed symptoms of eosinophilia-myalgia syndrome

[see Chapter 1]. Eosinophil counts did not consistent-
ly rise or fall, and there was no change in routine blood
cell counts and in findings from urinalysis, blood
chemistry studies, and electrocardiograms.

Let's review some of the side effects that can occur if
high dosages of 5-HTP are used.

Gastrointestinal

The most frequently mentioned side effects are gastrointestinal, and include nausea, loss of appetite, diarrhea,
cramps, stomach upset, gas, and vomiting. This stems from
5-HTP's conversion into serotonin, since too much serotonin
in the digestive system can induce these types of symptoms.
These side effects occur more commonly when blood levels
of 5-HTP rise above 10 mcg per 100 ml (Magnussen, 1982).
Normal levels are about 0.1 mcg per 100 ml.

Giving intravenous 5-HTP at the rate of 0.8 mg per kilogram of body weight to depressed children resulted in 20
percent of the children becoming nauseated (Ryan, 1992).
This dose of 5-HTP would be equivalent to about 60 mg in
an average adult. I have experienced nausea after ingesting
100 mg of 5-HTP. My patients have rarely reported gastrointestinal symptoms when taking dosages of 50 mg or less.

Cardiovascular

Doctors have had experience in prescribing SSRIs for heart
patients, since depression is common in patients who have
suffered a heart attack. It is reassuring to know that SSRIs
have not been found to cause problems in patients with cardiovascular conditions, such as arrhythmias or coronary
artery disease (Wong, 1995). No adverse cardiovascular
effects are known to occur with proper dosages of 5-HTP. It

is possible, though, that if large numbers of people start using 5-HTP, we may discover previously unknown cardiovascular effects in a minority of patients.

Dermatological

Dermatomyositis, a painful skin and soft-tissue inflammation, and a scleroderma-like illness developed in one 70-year-old man treated with 1,400 mg of 5-HTP and 150 mg of carbidopa for a period of twenty months (Sternberg, 1980). After he discontinued both substances, his symptoms gradually improved.

Both sweating and cold feelings have been reported by patients who took more than 100 mg of 5-HTP a day.

Hormonal

Breast enlargement in both men and women has been reported with the long-term use of SSRIs, probably due to stimulation of the hormone prolactin (Amsterdam, 1997). The administration of 5-HTP could also stimulate prolactin release, but there have not been any reports in the medical literature of breast enlargement with 5-HTP. This is a possibility to keep in mind, though, in anyone who is planning to take this serotonin precursor for prolonged periods.

Neurological

Insomnia is possible in certain individuals who could have a paradoxical reaction to 5-HTP. Instead of 5-HTP use leading to sleepiness at night, some people may find that it causes alertness.

Tremors, seizures, confusion, and delirium have occurred after the sudden discontinuation of 5-HTP in patients who had been taking several thousand milligrams for one month (Wyatt, 1973).

Sexual

As a rule, elevation of serotonin levels decreases sexual drive, while elevation of dopamine levels has the opposite effect. Therefore, serotonin excess could certainly decrease interest in sex or interfere with enjoyment. In fact, one of the most frequent side effects of SSRIs is interference with sexual enjoyment. Excess use of melatonin also decreases interest in sex.

However, sometimes there's a paradoxical effect from medicines that increase serotonin levels. If a person is anxious and stressed because of serotonin deficiency, they may not have much of an interest in engaging in a sexual relationship. When their serotonin levels become more balanced, and they relax, they may then be more interested in becoming intimate.

If you find 5-HTP is interfering with your sex drive or interest, you can avoid taking it on days when you expect to engage in sex. For instance, if Saturday night is your big night, you can skip your dose on Friday and Saturday, and resume taking your dose on Sunday.

Miscellaneous

Fatigue, stuffy or runny nose, and headaches are infrequently mentioned side effects.

A case study of an adverse reaction

Doctors in Delhi, India, reported in the *American Journal of Psychiatry* a case of an adverse reaction experienced by a 30-year-old man. He had been diagnosed with depression, and after giving informed consent, was included in a study in which he was given oral 5-HTP. His past medical history included chronic atopic dermatitis, a skin condition, but there was no history of drug reaction. His laboratory studies were all normal.

After an overnight fast, Mr. A received 5-HTP at a dose of 3 mg per kilogram of body weight followed by a light breakfast. Approximately ninety minutes after taking the 5-HTP, he experienced itching and became restless. Soon he complained of dizziness, blurred vision, and difficult breathing. His pulse increased from 80 to 100 beats per minute, but there was no change in blood pressure. He was given cyproheptadine, a medicine that blocks serotonin receptors, and within two hours he had completely recovered.

The report does not mention the weight of the patient. Assuming he was of normal weight, about 155 pounds, his dose would have been over 200 mg of 5-HTP. I always recommend everyone start 5-HTP at a dosage not exceeding 25 or 50 mg.

Interactions with other therapies

Formal studies on the use of 5-HTP when combined with other supplements have not been published. Therefore, it is difficult to make recommendations regarding combinations with great certainty. The information in this section relies mostly on my knowledge of nutrition and medicine, my personal experience with 5-HTP, reports from patients, interviews with users of 5-HTP, and discussions with physicians who use this nutrient in their practices. As such, I want to emphasize that the following recommendations are based on preliminary information.

Vitamins

There should be no problems in using 5-HTP along with your multivitamins. However, one possible area to consider is the use of 5-HTP concurrently with vitamin B_6, or pyridoxine. Vitamin B_6 is a cofactor involved in converting 5-HTP to serotonin. See Chapter 11 for a discussion of this topic.

One caution I would have is to make sure you're not taking too high a dosage of the B vitamins. Taking between two to five times the Recommended Daily Allowance (RDA) should be sufficient. Some tablets may contain from 50 to 100 times the RDA. Very large dosages may cause an imbalance, since the proportion of each specific vitamin in most B-complex tablets is not balanced. For instance, thiamine, or vitamin B_1, is cheap and biotin is expensive. You may be getting many times the RDA of vitamin B_1, but only a fraction of the RDA for biotin.

Minerals

No known interactions can be foreseen if 5-HTP is used along with minerals such as calcium, magnesium, zinc, selenium, and others.

Nutrients

5-HTP should not interfere with carnitine, creatine, glucosamine, fish oils, acidophilus, flaxseed oil, grape seed extract, lecithin, choline, carotenoids, flavonoids, pine bark extract, or soy extracts.

Herbs

St. John's wort can be used in combination with 5-HTP when both are used in small dosages, although one needs to be aware that high dosages could induce SS. (See Chapter 6 for details.) Ginkgo biloba is known to improve memory. It can cause slight alertness and should not interfere with 5-HTP. Ginseng, ashwagandha, and other herbs that have a mild stimulant effect should not interfere with the use of 5-HTP, except in those who have anxiety. Sometimes stimulants can exacerbate tension and irritability. Many of these herbs may work well with 5-HTP, since they may provide a certain level of alertness that counter-act 5-HTP's sedative properties.

Saw palmetto, garlic, echinacea, goldenseal, and other herbs that don't have a noticeable effect on the mind should not create problems when taken in conjunction with 5-HTP.

Hormones

There is no reason to think that 5-HTP interferes with the use of estrogen replacement therapy. The hormones DHEA, pregnenolone, and androstenedione cause alertness. When taken during the day, they would partially prevent the seda-tion that can result from 5-HTP use. High doses of these hor-mones can lead to anxiety and irritability. (For more infor-mation, see *DHEA: A Practical Guide* and *Pregnenolone: Nature's Feel Good Hormone* in Appendix B.)

Medicines

In Chapters 5 through 7, I discussed the use of 5-HTP in conjunction with pharmaceutical antidepressants, antianx-iety agents, and sleeping pills. I believe it is possible that some patients may benefit from these combinations if this is done cautiously and under medical supervision.

5-HTP, in small dosages, does not seem to have a sig-nificant effect on the cardiovascular system or the heart's electrical activity. Nevertheless, be careful if you are cur-rently on heart medicines or blood pressure pills.

There are countless classes of medicines, such as antibi-otics, narcotics, ulcer medicines, bronchodilators, and mus-cle relaxers, just to name a few. As a rule, if you're planning to add 5-HTP to your regimen, make sure your health care provider is aware of it and plans to supervise you. Always start with low dosages of supplements whenever you intend to add them to your existing medicine regimen. In the case of 5-HTP, starting with 10 mg would be a cautious approach if you are currently being treated with one or more drugs.

Medical evaluations while you're on 5-HTP

Until we learn more about the short- and long-term effects of 5-HTP, I would recommend being closely evaluated by a health care provider if you are using this nutrient regularly for prolonged periods. He or she may wish to monitor the following:

- Vital signs, such as blood pressure, heart rate, and temperature.

- Weight, since 5-HTP can cause appetite suppression.

- Mood, motivation, alertness, and energy levels.

- Sleep patterns, the time it takes to fall asleep, level of alertness or grogginess in the morning, number of times of awakenings throughout the night, intensity of dreams, and length of sleep.

- Routine blood studies, such as blood count, liver enzymes, and kidney function. These could be done every three to four months.

- Although no studies have indicated that using 5-HTP causes any heart valve abnormalities (see page 136), I would recommend anyone who is planning to take 5-HTP continuously, especially in dosages greater than 75 mg a day, to have an echocardiograph done every few months to evaluate the heart valves.

Summary

Please realize that there is no indication at this time that taking 5-HTP will cause any major problems. But, as with any medicine or supplement that has not had extensive testing, it's best to be cautious until more information becomes available. Sometimes we learn about the side effects of medicines and nutrients after they have been introduced to the general public. It is always wise to learn as much as you can about a supplement before taking it.

SUMMARY

5-HTP, the serotonin precursor, shows tremendous potential in the therapy of a variety of medical and psychiatric conditions. Here's a brief recount of its possible uses:

Appetite control—When used on a temporary basis, 5-HTP can help reduce food cravings and help you lose weight while you are trying to develop nonpill approaches to weight loss.

Depression—5-HTP can elevate mood by itself and is beneficial for some people who have mild depression. If you have moderate depression, you would need to combine 5-HTP with other herbs, nutrients, and medicines.

Anxiety—5-HTP can help you relax and feel calmer. Excessive daytime dosages can lead to sedation and the urge to nap. High-dose daytime use is not recommended if you plan to drive or operate heavy machinery.

Insomnia—It is most likely that 5-HTP will help you fall and stay asleep. A minority of patients finds 5-HTP to interfere with sleep, since it can paradoxically lead to alertness.

Tolerance could develop with nightly use, and I don't recommend relying on 5-HTP as a sleep aid on a regular basis. Using it once or twice a week should not lead to tolerance. Side effects when high dosages are used include vivid dreaming.

Fibromyalgia—Although only partially effective by itself, 5-HTP may allow you to sleep more deeply—and that's part of the battle in treating this disorder. Regular nightly use of 5-HTP is not recommended.

PMS—5-HTP can help you relax, and provide a sense of calmness and slight mood elevation at a time of the month when you need it most.

Migraine headaches—Limited studies have shown that 5-HTP decreases the severity of migraines. More research is certainly needed.

Final commentary

My clinical experience, and that of the physicians I interviewed for this book, does not indicate 5-HTP to have any significant side effects when used in low dosages. Until we learn more, I would recommend limiting your continuous use to less than three months, and to limit daily dosages to 75 mg. There may be occasions when a higher dosage would be temporarily required. As with any medicine, it's best to use the lowest effective dosage, and to occasionally take breaks. While you're taking these breaks, if your medical or psychiatric condition requires it, you could substitute other nutrients, herbs, or medicines. After a break for two to four weeks, you could resume taking the 5-HTP.

If you wish to keep up with the latest research on 5-HTP and other supplements, see the back of this book on how to subscribe to *Longevity Research Update*. Also see my regularly updated website at www.raysahelian.com.

A NOTE TO PROFESSIONALS

I have provided this appendix for those of you who are interested in the more detailed aspects of 5-HTP and serotonin research.

Where in the brain is 5-HTP?

Researchers at the Department of Psychiatry, University Hospital in Uppsala, Sweden, have determined through positron emission tomography (PET) scanning that 5-HTP and serotonin are present in most parts of the brain, particularly in the frontal cortex and striatum (Reibring, 1992).

What are normal blood levels of 5-HTP?

The estimated normal blood level of 5-HTP is about 0.1 microgram per 100 milliliters (Young, 1982). One gram equals 1,000 milligrams, and one milligram equals 1,000 micrograms.

5-HTP or tryptophan?

I'm often asked about the differences between 5-HTP and tryptophan. Both have similar effects, but there are some variances in their physiological effects. These include:

● Tryptophan is not absorbed as well from the intestines, and much of the absorbed portion is metabolized by enzymes found in the liver and other parts of the body (Fernstrom, 1983).

● Tryptophan is converted into 5-HTP, but a portion can be degraded by an enzyme called tryptophan pyrrolase with the end product being chemicals with no known function, such as kynurenine and quinolinic acid (Zmilacher, 1988). Tryptophan can also be used to make the vitamin niacin.

● As Dr. W. Poldinger, from the Psychiatric University Hospital in Basel, Switzerland, says, "The results obtained with tryptophan in depression are somewhat ambiguous. Its metabolic fate may send it to any of several pathways, and only one of these leads to serotonin. Moreover, for entry into the brain, tryptophan is dependent upon a transport system which it shares with five other amino acids, the consequence being competitive inhibition." (Poldinger, 1991)

● The rate-limiting step in the formation of serotonin is the enzyme tryptophan hydroxylase. By providing 5-HTP, we bypass this rate-limiting step and can produce serotonin in significant amounts.

● 5-HTP easily crosses the blood-brain barrier since, during its circulation in the blood, it is not bound to a carrier protein such as albumin (Zmilacher, 1988).

● In a double-blind comparative study, tryptophan was

not as effective as an antidepressant as was 5-HTP (van Praag, 1986).

Do brain 5-HTP levels stay constant throughout life?

A Japanese study done in 1992 on 144 subjects showed that, for some unknown reason, the levels of 5-HTP in the cerebrospinal fluid bathing the brain tends to increase with age (Tohgi, 1993). The same was true of several other brain chemicals, including tyrosine, dopamine, and norepinephrine. The researchers state, "These findings suggest that the concentrations of monoamine and amino acid transmitters and their related compounds in the cerebrospinal fluid reflect age-related changes in the synthesis, release, and reuptake mechanisms of the transmitters and their transport mechanisms across the blood-brain barrier."

The clinical significance of these findings is not clear.

Serotonin as an antioxidant?

There is a possibility that serotonin could even have antioxidant capabilities (Huether, 1996). This means it may fight the actions of free radicals, harmful substances that result from oxygen use. Dr. G. Huether, from the University of Göttingen in Germany, reports, "Serotonin is stored in platelets and is released in emergencies in large quantities at sites of irritation. Because such sites are always areas of increased [free] radical formation, the primary function of the platelet-stored 5-HT may be that of a unique, i.e., locally releasable, antioxidant and radical scavenger." More research is needed to determine whether serotonin is an antioxidant, and whether the ingestion of 5-HTP, with its immediate conversion into serotonin, would provide us with additional antioxidant protection.

Manipulating our neurotransmitters

We are in the age of modern science in which we can manipulate, to some extent, our moods, thoughts, and behaviors through the ingestion of nutrients, hormones, and medicines. The moral and ethical implications of this self-manipulation, and the role that government should or should not play in restricting our access to these medicines, are beyond the scope of this book. My intent as a scientist is to just give you the facts and let you decide for yourself how you wish to use this information.

Exercise, travel, falling in love, meditation, Yoga, relaxation techniques, massage, acupuncture, choice and timing of meals—these are just a few natural, nonspecific methods we can use to manipulate our brain chemicals and make ourselves feel better. However, there are several pharmaceutical ways serotonin levels in the brain can be precisely maneuvered:

- By stimulating additional serotonin production, for instance, through provision of the precursors tryptophan and 5-HTP.

- By inhibiting its formation, for instance, through use of the toxic drug p-chlorophenylalanine.

- By preventing its reuptake back into the neuron that released it in the first place, for instance, through use of the selective serotonin reuptake inhibitors (SSRIs).

- By inhibiting the enzymes, called monoamine oxidases (MAOs), that break it down.

- By providing chemicals that stimulate its release from the neuronal vesicles, such as MDMA (the drug Ecstasy) and Pondimin (fenfluramine), the weight-loss pill.

- By providing chemicals that mimic its actions at its various receptors, such as the antianxiety agent Buspar (bus-pirone).

- By providing chemicals that inhibit its actions at various receptors, such as the serotonin receptor blocker Zofran (ondasetron) used to prevent nausea and vomiting during chemotherapy.

Thus far, I've discussed serotonin receptors as if there were only just one type. But our brains are more complicated than we think and there are a number of types of serotonin receptors.

The serotonin receptor family

Serotonin influences brain chemistry by acting directly on receptors located on brain cells, or neurons. However, most people don't realize that there is quite a variety of serotonin receptor types in the brain. In the past decade, there has been tremendous progress in identifying serotonin receptors. It now appears that there are at least seven types of neuron receptors for serotonin: 5-HT1, 5-HT2, 5-HT3, 5-HT4, 5-HT5, 5-HT6, and 5-HT7. "HT" stands for hydroxytryptamine, the chemical name for serotonin.

Each of these types has their own subtypes. For instance, 5-HT1 can be subclassified as 5-HT1A, 5-HT1B, 5-HT1C, and so on. Scientists have been able to develop specific chemicals that can almost exclusively influence any one subtype. When we provide 5-HTP, this converts to serotonin, and all the receptors, no matter what type, may be influenced.

Here are some specific examples of serotonin receptor subtypes being influenced by specific medicines. The word "agonist" means that the drug is able to combine with the receptor and enhance its action, while "antagonist" means that the drug can block the receptor and prevent its action:

- Buspar (buspirone) is an agonist at 5-HT1A receptors and works as an antianxiety agent.

- Imitrex (sumatriptan) is a 5-HT1D receptor agonist for the treatment of acute migraine attacks.

- The administration of an investigational drug known as m-chlorophenylpiperazine decreases appetite through the specific activation of 5-HT2C receptors in the hypothalamus (Sargent, 1997).

Monoamine oxidase A and B

There are two types of MAO, types A and B. MAO-A preferentially deaminates serotonin, epinephrine, and norepinephrine, and is found mainly in the intestine and liver (Martin, 1996). Deamination means breaking off an amine, or a nitrogen molecule attached to two hydrogen molecules. MAO-B preferentially deaminates phenylethylamines and phenylethanolamines, and is responsible for all MAO activity in platelets and 80 percent of the MAO activity in the brain. Both type A and type B MAO enzymes can also deaminate dopamine.

Inhibition of MAO-B is considered essential for antidepressant effects. MAOI drugs, such as Nardil, are nonselective, meaning they have both A and B activity. Deprenyl (selegiline), a medicine currently used for Parkinson's disease, is a MAO-B inhibitor.

The metabolism of serotonin

Scientists can get a good idea of the serotonin content in the central nervous system by measuring the amount of the metabolite, or breakdown product, of serotonin in spinal fluid, blood, or urine. This metabolite is known as 5-hydroxyindoleacetic acid (5-HIAA). Similarly, one can measure the amount of norepinephrine by measuring the levels of its metabolite, methoxy-hydroxyphenylglycol (MHFG).

Measuring these metabolites can give us an indication whether a person's depression is caused more by a deficiency of serotonin than a deficiency of norepinephrine. However, this is an overly simplistic evaluation of depression, since there are other brain chemicals and hormones that influence mood. And we also need to keep in mind that sometimes there's no good correlation between total levels of brain chemicals and mood—a shortage of serotonin does not necessarily lead to manifestations of depressive symptoms (Poldinger, 1991). There could be a shortage of serotonin in a specific area of the brain that could account for a significant behavioral or mood disorder. Furthermore, we need to consider the function and health of receptors on brain cells since they can be either tolerant, or supersensitive, to brain chemicals.

More on serotonin

Although scientists had suspected the existence of a substance such as serotonin in blood since the middle of the nineteenth century, it wasn't until the middle of this century that serotonin was finally isolated. Over the years, scientists have discovered these facts about serotonin:

- Only about 1 to 2 percent of the serotonin in the whole body is found in the brain.

- This substance is found in large amounts in platelets, small cells in the blood that help form clots; mast cells, immune cells outside the bloodstream that are involved in allergies; and enterochromaffin cells, cells scattered in the digestive tract that produce such hormones as gastrin, glucagon, and others.

- Serotonin depletion is known to cause hypersexuality (Sheard, 1969). When P-chlorophenylalanine, a toxic medicine that depletes serotonin, is given to animals, there's a release from inhibitions with a dramatic increase in sexual behavior. Females start behaving like males; animals of both sexes start mounting other animals of the same sex. They even mount different species.

- The fact that serotonin is not found in any significant quantities in the cardiac system explains why SSRIs that manipulate serotonin levels generally do not cause heart irregularities.

Serotonin's notorious cousins

A number of chemicals with structures similar to that of serotonin are also found in nature. Some of these possess powerful psychotropic activities, including the inducement of hallucinations, and of altered states of awareness and consciousness. These include the psychedelics, such as 5-hydroxy-N,N-dimethyltryptamine (bufotenine) and 4-phosyphoryloxy-N, N-dimethyl-tryptamine (psilocybin).

The influence of 5-HTP on hormones

It is widely accepted that serotonin plays a significant role in the regulation of several hormones in the brain and body. The healthy connection between the hypothalamus, pituitary gland, and the adrenal glands—called the HPA axis—is often evaluated by giving medicines that stimulate serotonin production or release. A healthy functioning hormonal system responds appropriately to serotonin by doing the following: the hypothalamus releases corticotropin-releasing hormone (CRH), which stimulates the pituitary gland to release adrenocorticotropic hormone (ACTH). ACTH, in turn, stimulates the secretion of cortisol, aldosterone, and androgenic steroids from the adrenal glands.

Serotonin and hormones in major depression

There is now evidence that major depression is accompanied by an increased activity of the HPA axis, along with disorders in brain and peripheral serotonin activity (Maes, 1996). This increased activity leads to excess cortisol production. One test that psychiatrists use to diagnose major depression is the overnight dexamethasone test. Dexamethasone is a powerful corticosteroid that suppresses the hypothalamus from releasing ACTH, the pituitary hormone that stimulates the release of cortisol from the adrenal glands. When dexamethasone is given to healthy individuals at night, it will suppress cortisol production. This is evaluated by testing blood levels of cortisol in the morning. However, individuals who are depressed have an overactive HPA system and the administration of dexamethasone is not able to suppress cortisol release.

Several studies have been done to evaluate the influence of 5-HTP on the release of hormones in the body and brain. Most of these studies have shown consistent results.

One of the hormones most commonly influenced by 5-HTP is prolactin. Prolactin, a hormone made of 198 amino acids, is made and secreted by the pituitary gland. Prolactin plays several roles in the body. After a mother delivers a baby, the release of prolactin stimulates breast tissue to secrete milk. High levels of prolactin can decrease the functioning of the gonads, and can lead to a slowing or stoppage of menstruation in women. Excessive release of prolactin is called hyperprolactinemia. If continued for prolonged periods, it can lead to infertility. In men, excessive prolactin release leads to decreased testosterone production and sperm formation. This presents clinically as impotence, decreased libido, and infertility. The effects are reversed upon stoppage.

The release of prolactin in the body is episodic. There's an increase noted an hour or two after sleep. Peak levels are found between 4 A.M. and 7 A.M.

The release of prolactin is influenced by levels of brain chemicals and by certain medicines. For instance, dopamine inhibits the release of prolactin. Therefore, medicines that increase dopamine levels, such as bromocriptine, will decrease the amount of prolactin released. In contrast, medicines that increase serotonin release will enhance prolactin secretion.

Let's review a few of the studies published on the role of 5-HTP, tryptophan, and serotonin in the release of brain hormones.

1982: tryptophan and hormone release

Tryptophan is known to stimulate a hormonal response similar to that of its precursor, 5-HTP. Ten healthy subjects were given intravenous tryptophan and had several hormone levels measured (Charney, 1982). The tryptophan infusion induced a significant increase in prolactin release in all ten subjects. An increase in growth hormone concentrations was also observed, but this response was more variable. The subjects reported feeling significantly more high, mellow, and drowsy following the tryptophan administration when compared with the placebo group.

1989: 5-HTP and prolactin release

This randomized, placebo-controlled crossover study was conducted at the Thomas Jefferson Medical School in Philadelphia (my alma mater). Eight healthy men between the ages of nineteen and forty-two who were already taking the corticosteroid dexamethasone received 100 mg of 5-HTP four times a day. The 5-HTP was given in conjunction with 50 mg of carbidopa, a peripheral decarboxylase inhibitor (PDI) (Vlasses, 1989). Serum prolactin levels increased in all the patients when compared with those who received placebos, but there was no increase in aldosterone secretion. The researchers state, "The increases in prolactin observed after oral 5-HTP/carbidopa in our study provide further evidence for a role of serotonin in the regulation of prolactin release in man."

1992: 5-HTP and hormone release in depressed children

Investigators at the Department of Psychiatry, University of Pittsburgh intravenously administered 0.8 mg per kilogram body weight of 5-HTP to thirty-seven depressed prepubertal children, and compared the results with those found in children who were not depressed. Seven of the thirty-seven children had nausea or vomiting and did not complete the study. There was no change in the amount of growth hormone released (Ryan, 1992). The depressed children secreted less cortisol and more prolactin. The prolactin release occurred more in the depressed girls. The researchers say, "These studies are consistent with dysregulation of central serotonergic systems in childhood major depression."

1997: 5-HTP, hormone release, and the menstrual cycle

Twenty-two healthy women between the ages of eighteen and twenty-five were given intravenous 5-HTP (Lado-Abeal, 1997). Levels of leutenizing hormone (LH) increased when the women were given 5-HTP during the first part of the menstrual cycle (follicular phase). However, 5-HTP did not stimulate the release of LH during the second part of the menstrual cycle (luteal phase).

Bulimia nervosa, serotonin, and hormone levels

Bulimia is defined as the regular uncontrolled ingestion of large amounts of food followed by attempts to drastically lose the weight thus gained through diuretics, laxatives, strict dieting, or self-induced vomiting. It occurs most commonly in young, white middle- and upper-class women. This eating disorder was first identified in 1979 (Russell).

Several studies have suggested that bulimia is caused by a combination of genetic predisposition acting in concert with life stresses that decrease serotonin function in the brain. In recent years, evidence has accumulated that disturbances in brain serotonin are involved in this condition (Goldbloom, 1990; Jimerson, 1997). There is an area in the brain called the hypothalamus that is involved in eating behavior. The hypothalamus has a number of serotonin pathways that participate in influencing hunger and satiety. It has been shown that premeal administration of substances that increase serotonin in the brain, such as tryptophan and chlorophenylpiperazine, lead to decreased food intake (Brewerton, 1994).

In comparison with healthy individuals, bulimia patients don't feel as full after eating a meal (LaChaussee, 1992). A further line of evidence relating serotonin to bulimia is that diverse disorders, such as depression, anxiety, suicide proneness, alcoholism, aggression, and impulsivity, frequently co-exist with bulimia (Van Praag, 1987; Coccaro, 1989).

Researchers from the University of Toronto wished to investigate the role of serotonin in bulimia nervosa (Goldbloom, 1996). The immediate precursor to serotonin, 5-HTP, was used to stimulate serotonin synthesis and to see what effect this would have on prolactin, growth hormone, and cortisol levels. The release of these hormones is, in part, influenced by serotonin.

Eight female patients received an intravenous infusion of 5-HTP at a dose of 0.4 mg per kilogram of body weight. The results were compared with those of a control group without bulimia who received the same dose. Unlike the control group, the patients with bulimia did not show an increase in prolactin or growth hormone release in response to the 5-HTP, but showed a higher level of cortisol release. This study indicates that patients with bulimia have a disturbance in serotonin function. Whether 5-HTP supplementation is useful in these patients has currently not been determined, but is certainly worth a trial.

Summary of 5-HTP and hormone influences

The release of prolactin by 5-HTP leads one to wonder whether the continuous use of this nutrient could lead to a decrease in libido. It is unclear at this time whether there's an adaptation of the body with time to 5-HTP stimulation of prolactin. No long-term studies have been done to evaluate the effects of regular use of 5-HTP on the hypothalamic-pituitary axis and other hormonal systems.

Appendix B

Suggested Reading Materials

The Brain Wellness Plan. Jay L. Lombard, C. Germano. Kensington Publishing, New York, 1997.

DHEA: A Practical Guide. Ray Sahelian. Avery Publishing Group, Garden City Park NY, 1996.

Fibromyalgia: A Comprehensive Approach. Miryam Williamson. Walker and Co., New York, 1998.

From Fatigued to Fantastic! Jacob Teitelbaum. Avery Publishing Group, Garden City Park NY, 1996.

Kava: The Miracle Antianxiety Herb. Ray Sahelian. St. Martin's Press, New York, 1998.

Melatonin: Nature's Sleeping Pill. Ray Sahelian. Avery Publishing Group, Garden City Park NY, 1995.

Pregnenolone: Nature's Feel Good Hormone. Ray Sahelian. Avery Publishing Group, Garden City Park NY, 1997.

St. John's Wort: Nature's Feel-Good Herb. Ray Sahelian. Impakt Communications, Green Bay WI, 1997.

The Way Up From Down. Priscilla Slagle. St. Martin's Press, New York, 1988.

PERSONAL STORIES

E ver since the introduction of 5-HTP to the public, a number of people have taken it. I've gathered many anecdotes from individuals and some of my patients who have had a chance to try this nutrient. As you will note, not everybody has the same reaction to 5-HTP. Here are some of their experiences.

The first day I tried 5-HTP I was skeptical. Skeptical because almost anything I have tried previously from a vitamin store didn't affect me in any way. I guess I have a high threshold level. Nevertheless, a friend recommended I give 5-HTP a try. I took a 50-mg capsule in the early afternoon on an empty stomach at work, and, as I expected, I did not feel anything. That evening, on my drive home, I noticed a slight mood enhancement. It was strange, unlike anything I had felt before. I just wanted to smile. I was driving in Los Angeles traffic with a smile on my face—a rare site to see on the LA freeways at rush hour. That night I went out to visit a friend and again felt the same happy feeling I had experienced earlier. I thought the 5-HTP was still influencing me, but I was not yet totally convinced. Maybe it was just a coincidence that I was having a good day.

The next day I thought I should try it again and see if I would feel the same effect, and I took a 50-mg pill once again on an empty stomach. An hour later, I felt it. It was a relaxing feeling, without feeling tired. It was great, 5-HTP was working like an antistress medication. I thought to myself, "Wow, this nutrient is going to be big news; really big if even I can feel it." I can't imagine what it can do to others who are more sensitive to medications. That entire day I felt relaxed, yet I was quite productive.

Since those early days, I have taken 5-HTP in a 50-mg dose off and on for a few weeks now. I definitely notice the difference in mood on the days I take it. I feel more social, more relaxed, and the simple things that bring pleasure, like listening to music, feel even more pleasurable. I have never had trouble with sleep so I can't say if it has improved my sleep or not.

Ron, 25, Van Nuys, California

I have not taken 5-HTP for a couple of months now, and have not had many lucid dreams. I did have one about a week ago, though. The frequency of these lucid dreams is nowhere near the every-other-night ones I had with 5-HTP. I wish scientists would hurry up and prove this nutrient to be safe so I would feel more comfortable in taking it regularly.

David, 28, Lincoln, Nebraska

For about a year and a half, each morning I took a combination of from 50 to 100 mg of 5-HTP, along with 500 mg of phenylalanine, 18 mg of phentermine (a pharmaceutical antiobesity drug), 400 units of vitamin E, 500 mg of vitamin C, multivitamins, and fiber. I weighed 210 pounds before the start of this regimen, and now I'm down to 122 pounds.

I was monitored by a physician, and had physical exams, but did not have blood chemistry testing or electrocardiograms. I was overstimulated when my dosage of phentermine was greater than 26 mg, but I did not have side effects when I reduced the dosage to 18 mg. Taking the 5-HTP did not make me sleepy, since I was taking it with alertness-causing medicines.

Barbara Hirsch, Alexandria, Virginia

5-HTP makes me sleepy, but sometimes I feel alert. It seems that the sleepiness and alertness come and go in a cyclic fashion.

Van, 43, Newport Beach, California

I get a sense of well-being from 5-HTP, with a slight visual enhancement and a feeling of relaxation. However, I've noticed that if I take it all the time, I seem to build some tolerance to it.

Marina, 39, Dallas, Texas

Of all the natural supplements that I've tried, 5-HTP is the best when it comes to appetite control. I'm losing weight without even thinking about it.

Marge, 42, Brentwood, California

I use 5-HTP for sleep and it works well. I've played around with dosages from around 50 to 250 mg. The high dosage made me feel like a zombie the next day.

David Blanco, Grover Beach, California

I have four children to take care of and sometimes they drive me up the walls. If I alternate using 5-HTP and kava, I can keep the use of my Xanax to a minimum.

Sandy, 31, Tampa, Florida

Even though I've been taking 5-HTP only in the mornings, I've noticed my dreams to be more intense.

Jason, 34, New York, New York

I've been taking 50 mg of 5-HTP almost every night for the past two months. I definitely sleep better. I even put my eighty-year-old dad on it and he reports a deeper sleep on 5-HTP than he's had for over twenty years. He likes it more than melatonin. He has prostate enlargement and still gets up three times every night, but falls back asleep right way.

I certainly dream more with 5-HTP. I've had no side effects so far. Perhaps there's been a little tolerance building up, but no withdrawal symptoms on the nights that I don't take it, and no morning grogginess. My wife says I'm not a restless sleeper anymore.

Jeffrey, Omaha, Nebraska

I've tried 5-HTP on three separate occasions and didn't notice any effects.

JD, 27, Green Bay, Wisconsin

I've heard that 5-HTP is supposed to make you sleepy, but when I take it, I feel wired.

Marlene, 35, Des Moines, Iowa

I took 100 mg of 5-HTP at 3 P.M., and by 4 P.M. I was feeling sleepy and sluggish. I had an exercise class scheduled for the evening, and it was a struggle. My motivation was low.

Betty, 28, Beverly Hills, California

I took 5-HTP for one month. It was very helpful with no side effects. The only problem I had with it was the price, which was $40 for a month's supply. I used to take Prozac and later Paxil, which are paid for by my insurance. I'm probably going to continue the 5-HTP no matter what the cost because I can't stand the side effects of Paxil or Prozac.

Nancy, 29, Pittsburgh, Pennsylvania

I tried 5-HTP for a few weeks. The price is a killer. It makes me drowsy, and sometimes actually more depressed. I've gone back to my small dose of Prozac (10 mg) and have felt much better.

John, 46, New York, New York

I have taken tryptophan for many years at a dose of 2,000 mg nightly with vitamin B_6 and other cofactors, and did not notice the mood elevation as much as when I used 5-HTP. I've been on 5-HTP for a year and a half, generally at a dose of 50 mg most nights, with no side effects. I take it about a half-hour before bed. Sometimes I also take it early in the evening and this helps with sleep, too.

Will Block, Petaluma, California

After taking 60 mg of 5-HTP for a little over a week, an hour before bed (along with my usual dose of 3 mg of Xanax), I'm beginning to feel that I'm moving in a positive direction. Provided I get sufficient sleep, I'm feeling significantly calmer and clearer the next day than was typical without the 5-HTP. And most importantly, without the splitting headaches (and other side effects) that invariably accompanied the antidepressants I was prescribed.

RM, Atlanta, Georgia

We suspected our ten-year-old son had OCD. It began in June of 1997 and started getting steadily worse. He was engaged in more and more ritualistic actions, and his mood was dropping.

My wife and I began a search for a therapist with experience in OCD in children. Then one day I read an article about a supplement called 5-HTP. Within five days after my son started using 25 mg once a day, he showed some improvement. His obsessions were under control. He still has them, but they aren't nearly as debilitating or as strong. We still see him washing his hands, though not as much as before. It's been two months and he is progressing steadily, and we have not yet come across any side effects.

JT, San Diego, California

GLOSSARY

AADC (L-aromatic amino acid decarboxylase)—an enzyme that converts 5-HTP into serotonin. It also converts L-dopa into dopamine.

ACTH (adrenocorticotrophic hormone)—a hormone secreted by the pituitary gland. It stimulates the adrenal glands to make certain steroids, such as cortisol.

Amino acid—a molecule that serves as a unit of structure of proteins and contains nitrogen. Twenty-two amino acids are found in the human body, including arginine, lysine, tryptophan, and phenylalanine. Eight out of the twenty-two are essential, that is, the body cannot make them. They need to be ingested through foods.

Antioxidant—a substance that neutralizes damaging toxins, and thus prevents the deterioration of DNA, RNA, lipids, and proteins. Vitamins C and E, as well as beta-carotene, are the best known antioxidants, but more and more are being discovered each year.

Benzodiazepine—a class of medicines, such as Valium, that act on GABA receptors to induce relaxation and sleep.

Blood-brain barrier—the natural defense mechanism that restricts chemical access to the brain. It works by forcing various chemicals to compete for space on carrier proteins, which carry the chemicals from the bloodstream to the brain.

Control—in any study, whenever a group of animals or humans is given a certain medicine, it is compared with another group of animals or humans who are under the same circumstances for everything except the medicine. This second group is known as the control group. This way, researchers can find out the role of the medicine, independent of any other factors.

Cortisol—a substance secreted by the human adrenal glands. High levels lead to interference with the proper functioning of the immune system.

DHEA (dehydroepiandrosterone)—a hormone produced by the adrenal glands. DHEA, the most abundant adrenal hormone, is converted into male and female sex hormones.

Dopamine—a neurotransmitter that relays messages between neurons. It is one of the primary mood and alertness neurotransmitters.

Enzyme—a substance that causes, or speeds up, a chemical reaction within the body.

Eosinophilia-myalgia syndrome (EMS)—a disease that is marked by muscle pain and by the presence of many eosinophils, white blood cells involved in allergic reactions, in the blood. A number of people developed EMS after taking a contaminated tryptophan product, leading to an FDA ban on tryptophan in 1989.

GABA (gamma-amino-butyric acid)—a neurotransmitter that causes relaxation and sedation.

Gonad—a testicle or ovary.

Hormone—a chemical messenger produced by a gland or organ that influences a number of metabolic actions in the cells.

Hypothalamus—a small area of the brain above and behind the roof of the mouth. The hypothalamus is prominently involved with the functions of the autonomic nervous system and the hormonal system. It also plays a role in mood and motivation.

Melatonin—the body's primary sleep-inducing hormone, produced by the pineal gland. Serotonin can be converted into melatonin.

Metabolism—the chemical and physical processes continuously going on in the body involving the creation and breakdown of molecules.

Monoamine oxidase—an enzyme in the brain that breaks down neurotransmitters, such as serotonin.

Monotherapy—treatment with only one drug or supplement.

Neuron—a brain cell. There are over 100 billion of these cells in the human brain. Neurons communicate with each other through chemicals called neurotransmitters.

Norepinephrine—a neurotransmitter that relays messages between neurons. It is one of the primary mood and alertness neurotransmitters.

Peripheral decarboxylase inhibitor (PDI)—a class of medicines that blocks the enzyme AADC from converting 5-HTP into serotonin outside of the brain.

Phenylalanine—an amino acid ingested through protein foods. It is converted into tyrosine, and then to the neurotransmitters dopamine and norepinephrine.

Pineal gland—a small gland, shaped somewhat like a pine cone, located in the middle of the brain. It secretes melatonin, which influences the sleep-wake cycle.

Placebo—a dummy pill that contains no active ingredient.

Polytherapy—treatment with more than one drug or supplement.

Precursor—a substance that precedes, and is the source of, another substance.

Pregnenolone—the steroid from which DHEA, progesterone, and other steroids are formed.

Progesterone—a female hormone that, among other functions, helps regulate the menstrual cycle.

Prolactin—a hormone that stimulates milk production in nursing mothers, and also influences gonad function.

Receptor—a special arrangement on a cell that recognizes a molecule and interacts with it. This allows the molecule to either enter the cell or stimulate it in a specific way.

Reuptake—the reabsorption of a neurotransmitter from the synaptic cleft by the neuron that released it in the first place. This process stops the neurotransmitter's activity.

Serotonin—one of the primary mood neurotransmitters. It is derived from 5-HTP.

Steroid hormone—a hormone, such as DHEA or estrogen, made in the outer layers of the adrenal glands, which sit atop the kidneys. These hormones are also made in the gonads and the brain.

Synaptic cleft—the space between two neurons. Neurotransmitters carry messages from one neuron to another across the cleft.

Tolerance—the resistance the body develops when a drug or supplement is used continuously over time. This tends to negate the effects of the drug or supplement.

Tryptophan—an amino acid that needs to be ingested through foods, since the body cannot manufacture it. It is converted into 5-HTP, which in turn is converted into serotonin.

Tryptophan hydroxylase—an enzyme that converts tryptophan into 5-HTP.

Tyrosine—an amino acid ingested through protein foods. It is converted into the neurotransmitters dopamine and norepinephrine.

REFERENCES

Albus M, Zahn TP, Breier A. *Anxiogenic properties of yohimbine. Behavioral, physiological and biochemical measures.* Eur Arch Psychiatry Clin Neurosci 241:337–344, 1992.

Amsterdam JD, Garcia-Espana F, Goodman D, Hooper M, Hornig-Rohan M. *Breast enlargement during chronic antidepressant therapy.* J Affective Disorders 46:151–156, 1997.

Angst J, Woggon B, Schoepf J. *The treatment of depression with 5-HTP versus imipramine. Results of two open and one double-blind study.* Arch Psychiatr Nervenkr 224(2):175–186, 1977.

Autret A, Minz M, Beillevaire T, Degos C, Cathala HP. *Clinical and polygraphic effects of d.l 5HTP on narcolepsy-cataplexy.* Biomedicine 27(5):200–203, 1977.

Ballenger JC, Wheadon DE, Steiner M, Bushnell W, Gergel IP. *Double-blind, fixed-dose, placebo-controlled study of paroxetine in the treatment of panic disorder.* Am J Psychiatry 155:36–42, 1998.

Bellivier F, Leboyer M, Couriet P, et al. *Association between the tryptophan hydroxylase gene and manic-depressive illness.* Arch Gen Psychiatry 55:33–37, 1998.

Belongia EA, Gleich GJ. *The Eosinophilia-myalgia syndrome revisited* (editorial). J of Rheumatology 23;10:1682–1684, 1996.

Bigelow L, Wallis P, Gillin JC, Wyatt RJ. *Clinical effects of 5-HTP administration in chronic schizophrenic patients.* Biological Psychiatry 14:53, 1979.

Birmaher B, Kaufman J, Brent DA, Dahl RE, Perel JM, al-Shabbout M, et al. *Neuroendocrine response to 5-hydroxy-L-tryptophan in prepubertal children at high risk of major depressive disorder.* Arch Gen Psychiatry 54(12):1113–1119, 1997.

Blier P, Bergeron R. *Sequential administration of augmentation strategies in treatment-resistant obsessive-compulsive disorder: preliminary findings.* Int Clin Psychopharmacol 11(1):37–44, 1996.

Borne, R. *Serotonin: the neurotransmitter for the '90s.* Drug Topics, p. 108, 10 October 1994.

Bourin M, Baker GB, Bradwejn J. *Neurobiology of panic disorder.* J Psychosomatic Research 44(1):163–180, 1998.

Bowden CL. *Role of newer medications for bipolar disorder.* J Clinical Psychopharmacology 16[suppl 1]:48S–55S, 1996.

Brewerton TD, Murphy DL, Jimerson DC. *Testmeal responses following m-chlorophenyliperazine and L-tryptophan in bulimics and controls.* Neuropsychopharmacology 11:63–74, 1994.

Brown P, et al. *Effectiveness of piracetam in cortical myoclonus.* Mov Disord 8:63, 1993.

Bushan B, Seth SD, Saxena S, Gupta YK, Karmarkar MG. *An adverse reaction to L-5-hydroxytryptophan* (letter). Am J Psychiatry 148(2):268–269, 1991.

Byerley WF, Judd LL, Reimherr FW, Grosser BI. *5-hydroxytryptophan: a review of its antidepressant efficacy and adverse effects.* J Clin Psychopharmacol 7(3):127–137, 1987.

Cangiano C, Ceci F, Cascino A, Del Ben M, Laviano A, Muscaritoli M, Antonucci F, Rossi-Fanelli F. *Eating behavior and adherence of dietary prescriptions in obese adult subjects treated with 5-hydroxytryptophan.* Am J Clin Nutr 56(5): 863–867, 1992.

Caruso I, Puttini PS, Cazzola M, Azzolini V. *Double-blind study of 5-hydroxytryptophan versus placebo in the treatment of primary fibromyalgia syndrome.* J International Medical Research 18:201–209, 1990.

Cavallo A, Richards GE, Meyer WJ, Waldrop RD. *Evaluation of 5-hydroxytryptophan as a test of pineal function in humans.* Hormone Research 27:69–73, 1987.

Ceci F, Cangiano C, Cairella M, Cascino A, Del Ben M, Muscaritoli M, Sibilia L, Rossi-Fanelli F. *The effects of oral 5-hydroxytryptophan administration on feeding behavior in obese adult female subjects.* J Neural Transm 76(2):109–117, 1989.

Charney DS, Heninger GR, Reinhard JF Jr, Sternberg DE, Hafstead KM. *The effects of intravenous L-tryptophan on prolactin and growth hormone and mood in healthy subjects.* Psychopharmacology (Berl) 77(3):217–222, 1982.

Coccaro EF, Siever LJ, Klar HM, Maurer G, Cochrane K, Cooper TB, Mohs RC, Davis KL. *Sertonergic studies in patients with affective and personality disorders: correlates with suicidal and impulsive aggressive behavior.* Arch Gen Psychiatry 46: 587–599, 1989.

Cook EH, Rowlett R, Jaselskis C, Leventhal BL. *Fluoxetine treatment of children and adults with autistic disorder and mental retardation.* J Am Acad Child Adolescent Psychiatry 31:739–745, 1992.

Cooper JR, Bloom FE, Roth RH. *The Biochemical Basis of Neuropharmacology.* 7th edition. New York: Oxford University Press, 1996.

Cornelius JR, et al. *Fluoxetine in depressed alcoholics.* Arch Gen Psych 54:700–705, 1997.

Crenshaw T, Godberg J. *Sexual Pharmacology: Drugs That Affect Sexual Function.* New York: Norton & Company, 1996.

Crofford LJ, Hallett M, et al. *An eosinophilia-myalgia syndrome related disorder associated with exposure to L-5-hydroxytryptophan.* J Rheumatol 21(12):2261–2265, 1994.

Davis L, Suris A, Lambert M, Heimberg C, Petty F. *Post-traumatic stress disorder and serotonin: new directions for research and treatment.* J Psychiatry Neurosci 22(5):318–326, 1997.

De Benedittis G, Massei R. *5-HT precursors in migraine prophylaxis. A double-blind cross-over study with L-5-hydroxytryptphan versus placebo.* Clin J Pain 3:123–129, 1986.

De La Torre JC, Mullan S. *A possible role for 5-hydroxytryptamine in drug-induced seizures.* J Pharm Pharmacol 22:858, 1970.

Den Boer JA, Westenberg HG. *Behavioral, neuroendocrine, and biochemical effects of 5-hydroxytryptophan administration in panic disorder.* Psychiatry Res 31(3):267–278, 1990.

DiLorenzo C, Williams CM, Hajnal P, Valenzuaela JE. *Pectin delays gastric emptying and increases satiety in obese subjects.* Gastroenterology 95:1211–1215, 1988.

Disertori B, Ducati A, Piazza M. *Spasmodic torticollis, substantiating Manto syndrome, of possible toxic etiology, with alterations of brainstem acoustic evoked potentials (BAEPs). Treatment with L-5-hydroxytryptophan. Follow up of 18 months, during which high degree of resolution of symptoms and normalization of BAEPs took place.* Ann Osp Maria Vittoria Torino 25(1–6):3–20, 1982.

Fallon B, Liebowitz MR, Hollander E, Schneier FR, et al. *The pharmacotherapy of moral or religious scrupulosity.* J Clin Psychiatry 51:517–521, 1990.

Fellows LE, Bell EA. *5-HTP, 5-HT, and tryptophan-5-hydroxylase in Griffonia simplicifolia.* Phytochemistry 9:2389–2396, 1970.

Fernstrom JD. *Role of precursor availability in control of monoamine biosynthesis in brain.* Physiol Rev 99:484–546, 1983.

George DT, Lindquist T, Rawlings RR, Eckardt MJ, Moss H, Mathis C, Martin PR, Linnoila M. *Pharmacologic maintenance of abstinence in patients with alcoholism: no efficacy of 5-hydroxytryptophan or levodopa.* Clin Pharmacol Ther 52(5):553–560, 1992.

Gerhard U, Linnenbrink N, Geroghiadou C, Hobi V. *Vigilance-decreasing effects of 2 plant-derived sedatives.* Schwiz Rundsch Med Prax 9;85(15):473–481, 1996.

Goldbloom DS, Garfinkel PE. *The serotonin hypothesis of bulimia nervosa: theory and evidence.* Can J Psychiatry 35: 741–744, 1990.

Goldbloom DS, Garfinkel PE, Katz R, Brown GM. *The hormonal response to intravenous 5-hydroxytryptophan in bulimia nervosa.* J Psychosom Res 40(3):289–297, 1996.

Haensel SM, Klem TM, et al. *Fluoxetine and premature ejaculation: a double-blind, cross-over, placebo-controlled study.* J Clin Psychopharmacology 18:72–77, 1998.

Hagan JJ, Hatcher JP, Slade PD. *The role of 5-HT1D and 5-HT1A receptors in mediating 5-hydroxytryptophan induced myoclonic jerks in guinea pigs.* Eur J Pharmacol 29;294(2–3): 743–751, 1995.

Hartmann E, Greenwald D. *Tryptophan and human sleep: an analysis of 43 studies. In Scholossberger HG, Kochen W, et al. (eds). Progress in Tryptophan and Serotonin Research.* Berlin, New York: DeGruyter W, pp. 297–304, 1987.

Holzl J, Godau P. *Receptor binding studies with Valeriana officinalis on the benzodiazepine receptor.* Planta Medica 55:642, 1989.

Houghton PJ. *The biological activity of valerian and related plants.* J Ethnopharmacol 22:121–142, 1988.

Huether G, et al. *Effect of tryptophan administration on circulating melatonin levels in chicks and rats. Evidence for stimulation of melatonin synthesis and release in gastrointestinal tract.* Life Sciences 51:945–953, 1992.

Huether G, Schuff-Warner P. *Platelet serotonin acts as a locally releasable antioxidant.* In Graziella, Allegri, Filippini, et al. (eds). *Recent Advances in Tryptophan Research.* New York: Plenum Press, 1996.

Irwin MR, Marder SR, Fuentenebro F, Yuwiler A. *L-5-hydroxytryptophan attenuates positive psychotic symptoms induced by D-amphetamine.* Psychiatry Res 22(4):283–289, 1987.

Jacobsen FM, Sack DA, Wehr TA, Rogers S, Rosenthal NE. *Neuroendocrine response to 5-hydroxytryptophan in seasonal affective disorder.* Arch Gen Psychiatry 44(12):1086–1091, 1987.

Jimerson DC, Wolfe BE, Metzger ED, Finkelstein DM, Cooper TB, Levine JM. *Decreased serotonin function in bulimia nervosa.* Arch Gen Psychiatry 54:529–534, 1997.

Kahn RS, Westenberg HG. *L-5-hydroxytryptophan in the treatment of anxiety disorders.* J Affect Disord 8(2):197–200, 1985.

Kaneko M, Kumashiro H, Takahashi Y, Hoshino Y. *5-HTP treatment and serum 5-HT after 5-HTP loading on depressed patients.* Neuropsychobiology 5(4):232–240, 1979.

Kirk M, Pace S. *Pearls, pitfalls, and updates in toxicology.* Emerg Med Clin North Am 15:427–449, 1997.

Knutson B, et al. *Selective alteration of personality and social behavior by serotonergic intervention.* Am J Psychiatry 55:3, March 1998.

LaChaussee JL, Kissileff HR, Walsh BT, Hadigan CM. *The single-item meal as a measure of binge-eating behavior in patients with bulimia nervosa.* Physiol Behav 51:593–600, 1992.

Lado-Abeal B, Rey C, et al. *5-HTP amplifies pulsatile secretion of LH in the follicular phase of normal women.* Clinical Endocrinology (Oxf) Nov;47(5):555–563, 1997.

Larisch R, et al. *In vivo evidence for the involvement of dopamine D2 receptors in striatum and anterior cingulate gyrus in major depression.* Neuroimage 5:251–260, 1997.

Leathwood PD, Chauffard F. *Aqueous extract of valerian reduces latency to fall asleep in man.* Planta Med (2):144–148, 1985.

Lepetit P, Touret M, Grange E, Gay N, Bobillier P. *Inhibition of methionine incorporation into brain proteins after the systemic administration of p-chlorophenylalanine and L-5-hydroxytryptophan.* Eur J Pharmacol 17;209(3):207–212, 1991.

Levitan R, et al. Archives of General Psychiatry 55:244–249, 1998.

Li ETS, Anderson GH. *Amino acids in the regulation of food intake.* Nutr Abs Rev Clin Nutr 53:169–181, 1983.

Lichensteiger W, Mutzner V, Langemann H. *Uptake of 5-hydroxytryptamine and 5-hydroxytryptophan by neurons of the central nervous system normally containing catecholamines.* J Neurochem 14:489–497, 1967.

Lieberman HR, Wurtman JJ, Chew B. *Changes in mood after carbohydrate consumption among obese individuals.* Am J Clin Nutr 44:772–778, 1986.

Linnoila M, Virkkunen M, et al. *Low cerebrospinal fluid 5-hydroxyindoleacetic acid concentration differentiates impulsive from nonimpulsive violent behavior.* Life Sci 33:2609–2614, 1983.

Linnoila M, Virkkunen M. *Aggression, suicidality, and serotonin.* J Clin Psychiatry 10:46–51, 1992.

Lipson AH, Earl JW, Wilchen B, Yu JS, O'Halloran M, Cotton RG. *Successful treatment of dihydropteridine reductase deficiency, with an interesting effect of 5-hydroxytryptophan deficiency on sleep patterns.* J Inherit Metab Dis 14(1):49–52, 1991.

Lucca A, Lucini V, Catalano M, Alfano M, Smeraldi E. *Plasma tryptophan to large neutral amino acids ratio and therapeutic response to a selective serotonin uptake inhibitor.* Neuropsychobiology 29:108–111, 1994.

Maes M, Van Gastel A, Ranjan R, Blockx P, Cosyns P, Meltzer HY, Desnyder R. *Stimulatory effects of L-5-hydroxytryptophan on postdexamethasone beta-endorphin levels in major depression.* Neuropsychopharmacology 15(4):340–348, 1996.

Magnussen I, Jensen TS, Rand JH, Van Woert MH. *Plasma accumulation of metabolism of orally administered single dose L-5-hydroxytryptophan in man.* Acta Pharmacol Toxicol (Copenh) 49(3):184–189, 1981.

Magnussen I, Van Woert MH. *Human pharmacokinetics of long-term 5-hydroxytryptophan combined with decarboxylase inhibitors.* Eur J Clin Pharmacol 23:81–86, 1982.

Maissen CP, Ludin HP. *Comparison of the effect of 5-hydroxytryptophan and propranolol in the interval treatment of migraine.* Schweiz Med Wochenschr 121(43):1585–1590, 1991.

Martin TG. *Serotonin syndrome.* Ann Emerg Med 28(5): 520–526, 1996.

Martinelli I, Mainini E, Mazzi C. *Effect of 5-hydroxytryptophan on the secretion of PRL, GH, TSH and cortisol in obesity.* Minerva Endocrinol 17(3):121–126, 1992.

Martinez B, Kasper S, Ruhrmann S, Möller HJ. *Hypericum in the treatment of seasonal affective disorders.* J Geriatric Psychiatry Neurol 7(suppl 1):S29–S33, 1994.

McDougle CJ, Naylor ST, Cohen DJ, Aghajanian GK, Heninger GR, Price LH. *Effects of tryptophan depletion in drug-free adults with autistic disorder.* Arch Gen Psychiatry 53:993–1000, 1996a.

McDougle CJ, Naylor ST, Cohen DJ, Volkmar FR, Heninger GR, Price LH. *A double-blind, placebo-controlled study of fluvoxamine in adults with autistic disorder.* Arch Gen Psychiatry 53:1001–1008, 1996b.

Mendlewicz J, Youdin MBH. *Antidepressant potentiation of 5-hydroxytryptophan by L-deprenyl in affective illness.* J Affective Disord 2:137–146, 1980.

Michelson D, Page SW, Casey R, Trucksess MW, Love LA, Milstien S, Wilson C, Massaquoi SG, Crofford LJ, Hallett M, et al. *An eosinophilia-myalgia syndrome related disorder associated with exposure to L-5-hydroxytryptophan.* J Rheumatol 21(12):2261–2265, 1994.

Mitsikostas DD, Gatzonis S, Thomas A, Ilias A. *Buspirone vs amitriptyline in the treatment of chronic tension-type headache.* Acta Neurol Scand 96:247–251, 1997.

Morgane, PJ. *Serotonin: twenty years later. Monoamine theory of sleep: the role of serotonin—a review.* Psychopharmacol Bull 17:13–17, 1981.

Muller WE, Schafer CS. *St. John's wort: In vitro study about hypericum extract, hypericin and kanpderol as antidepressants.* Deutsche Apotheker Zeitung 136:1015–1022, 1996.

Nakajima T. *Amine precursor amino acid therapy: from neurochemical basis to clinical aspects.* Neurochem Res 21(2): 251–258, 1996.

Nicolodi M, Sicuteri F. *Fibromyalgia and migraine, two faces of the same mechanism. Serotonin as the common clue for pathogenesis and therapy.* Adv Exp Med Biol 398:373–379, 1996.

Nolen WA, van de Putte JJ, Dijken WA, Kamp JS, Blansjaar BA, Kramer HJ, Haffmans J. *Treatment strategy in depression. MAO inhibitors in depression resistant to cyclic antidepressants: two controlled crossover studies with tranylcypromine versus L-5-hydroxytryptophan and nomifensine.* Acta Psychiatr Scand 78(6):676–683, 1988.

Pasman WJ, Saris WHM, Wauters MAJ, Westerterp-Plantenga MS. *Effect of one week of fiber supplementation on hunger and satiety ratings and energy intake.* Appetite 29: 77–87, 1997.

Perovic S, Muller WE. *Pharmacological profile of hypericum extract. Effect on serotonin uptake by postsynaptic receptors.* Arzneimittelforschung 45(11):1145–1148, 1995.

Poldinger W, Calanchini B, Schwarz W. *A functional-dimensional approach to depression: Serotonin deficiency as a target syndrome in a comparison of 5-hydroxytryptophan and fluvoxamine.* Psychopathology 24:53–81, 1991.

Pranzatelli MR, Tate E, Huang Y, Haas RH, Bodensteiner J, Ashwal S, Franz D. *Neuropharmacology of progressive myoclonus epilepsy: response to 5-hydroxy-L-tryptophan.* Epilepsia 36(8):783–791, 1995.

Puttini PS, Caruso I. *Primary fibromyalgia syndrome and 5-hydroxy-L-tryptophan: a 90-day open study.* J Int Med Res 20(2):182–189, 1992.

Ranen NG, Lipsey JR, Treisman G, Ross CA. *Sertraline in the treatment of severe aggressiveness in Huntington's disease.* J Neuropsychiatry 8:338–340, 1996.

Rapin K, Katzman R. *Neurobiology of autism.* Ann Neurol 43(1):7–14, 1998.

Rapport MM, Green AA, Page IH. *Serum vasoconstrictor (serotonin). IV: isolation and characterization.* J Biol Chem 176:1243–1251, 1948.

Rasmussen S, Hackett E, DuBoff E, Greist J, Halaris A, et al. *A 2-year study of sertraline in the treatment of obsessive-compulsive disorder.* International Clinical Psychopharmacology 12:309–316, 1997.

Reibring L, Agren H, Hartvig P, Tedroff J, Lundqvist H, Bjurling P, Kihlberg T, Langstrom B. *Uptake and utilization of [beta-11C]5-hydroxytryptophan in human brain studied by positron emission tomography.* Psychiatry Res 45(4):215–225, 1992.

Ronnback AP, Jarvholm B. *Successful use of a selective serotonin reuptake inhibitor in a patient with multiple chemical sensitivities.* Acta Psychiatr Sand 96:82–83, 1997.

Russell GFM. *Bulimia nervosa: an ominous variant of anorexia nervosa.* Psychol Med 9:420–448, 1979.

Ryan ND, Birmaher B, Perel JM, Dahl RE, Meyer V, al-Shabbout M, Iyengar S, Puig-Antich J. *Neuroendocrine response to L-5-hydroxytryptophan challenge in prepubertal major depression. Depressed vs normal children.* Arch Gen Psy-chiatry 49(11):843–851, 1992.

Sano I. *L-5-hydroxytryptophan therapy.* Folia Pscyhiatr Neurol Japan 26:7–17, 1974.

Sargent PA, Sharpley AL, Williams C, Goodall EM, Cowen PJ. *5-HT2C receptor activation decreases appetite and body weight in obese subjects.* Psychopharmacology 133:309–312, 1997.

Schulz H, Stolz C, Muller J. *The effect of valerian extract on sleep polygraphy in poor sleepers: a pilot study.* Pharmacopsy-chiatry 27(4):147–151, 1994.

Shapiro S. *Epidemiologic studies of the association of L-Tryptophan with the eosinophilia-myalgia syndrome: a critique.* J Rheumatology 23;S46:44–48, 1996.

Sheard MH. *The effect of p-chlorophenyl-alanine on behavior in rats: Relation to brain serotonin and 5-hydroxy-indoleacetic acid.* Brain Research 15:524–528, 1969.

Sigal LH, Chang DJ, Sloan V. *18 tender points and the "18-Wheeler" sign: clues to the diagnosis of fibromyalgia.* JAMA 279(6):434, 1998.

Smith B, Prockop DJ. *Central-nervous system effects of ingestion of L-tryptophan by normal subjects.* N Engl J Med 267:1338–1341, 1962.

Smith KA, Clifford EM, Hockney RA, et al. *Effect of trypto-phan depletion on mood in male and female volunteers: a pilot study.* Hum Psychopharmacol Clin Exp 12:111–117, 1997.

Soulairac A, et al. *Action du 5-hydroxytrytophane, precurseur de la serotonine, sur les troubles du sommeil.* Annales Medico-psychologiques 135:792–798, 1990.

Stachow A, Jablonska S, Skiendzielewska A. *5-hydroxy-trypt-amine and tryptamine pathways in scleroderma.* Br J Dermatol 97(2):147–154, 1977.

Steiner W, Fontaine R. *Toxic reaction following the combined administration of fluoxetine and L-trytophan: five case reports.* Biol Psychiatry 21:1067–1071, 1986.

Sternberg EM, Van Woert MH, Young SN, et al. *Tolerance and dependence to serotonin: a speculation.* Arch Gen Psychiatry 29:597–599, 1973.

Sternberg EM, Van Woert MH, Young SN, et al. *Development of a scleroderma-like illness during therapy with L-5-hydroxy-tryptophan and carbidopa.* N Engl J Med 303:782–787, 1980.

Stokes PE. *Fluoxetine: a five-year review.* Clin Ther 15:216–243, 1993.

Stone RA, Worsdell YM, Fuller RW, Barnes PJ. *Effects of 5-hydroxytryptamine and 5-hydroxytryptophan infusion on the human cough reflex.* J Appl Physiol 74(1):396–401, 1993. To test the effects of 5-HTP on the human cough reflex, eight male volunteers were tested at the Department of Thoracic Medicine, National Heart and Lung Institute in London. These men were given an inhalation of sodium bicarbonate, which causes a cough reflex, and also were exposed to cap-saicin, which induces a worse cough response. The volun-teers were then given intravenous serotonin or 5-HTP. Both serotonin and 5-HTP were able to reduce the rate of cough-ing from the sodium bicarbonate, but not from the cap-saicin. 5-HTP did not induce a rise in heart rate, respiratory rate, or blood pressure. The researchers state, "We conclude that serotonin exerts an inhibitory influence over the human cough reflex at peripheral and possibly central sites."

Sundblad C, Wikander I, Andersch B, Eriksson E. *A natu-ralistic study of paroxetine in premenstrual syndrome: efficacy and side-effects during ten cycles of treatment.* European Neuropsychopharmacology 7:201–206, 1997.

Thiede HM, Walper A. *Inhibition of MAO and COMT by hypericum extracts and hypericin.* J Geriatr Psychiatry Neurol, 7(Suppl 1):S54–S56, 1994. The MAO inhibiting fraction contained hypericins as well as flavonols, the COMT-inhibition fraction being mainly flavonols and xanthones.

Titus F, Davalos A, Alom J, Codina A. *5-Hydroxytryptophan versus methysergide in the prophylaxis of migraine. A randomized clinical trial.* Eur Neurol 25(2):327–329, 1986.

Tiwary CM, Ward JA, Jackson BA. *Effect of pectin on satiety in healthy US Army adults.* J Amer College of Nutrition 16:(5):423–428, 1997.

Tohgi H, Takahashi S, Abe T. *The effect of age on concentrations of monoamines, amino acids, and their related substances in the cerebrospinal fluid.* J Neural Transm Park Dis Dement Sect 5(3):215–226, 1993.

Toubro S, Astrup AV, Breum L, Quaade F. *Safety and efficacy of long-term treatment with ephedrine, caffeine and an ephedrine/caffeine mixture.* International J Obesity 17(suppl 1):S69–S72, 1993.

Trouillas P, Garde A, Robert JM, Renaud B, Adeleine P, Bard J, Brudon F. *Regression of the cerebellar syndrome under long-term administration of 5-HTP or the combination of 5-HTP and benserazide. 26 cases quantified and treated using computer methods.* Rev Neurol (Paris) 138(5):415–435, 1982.

Trouillas P, Serratrice G, Laplane D, Rascol A, Augustin P, Barroche G, Clanet M, Degos CF, Desnuelle C, Dumas R, et al. *Levorotatory form of 5-hydroxytryptophan in Friedreich's ataxia. Results of a double-blind drug-placebo cooperative study.* Arch Neurol 52(5):456–460, 1995.

Turnbull WH, Thomas HG. *The effect of Plantago ovata seed-containing preparation on appetite variables, nutrient and energy intake.* Int J Obesity 19:338–342, 1995.

Uhde TW, Boulanger JP, et al. *Caffeine: relationship to human anxiety, plasma MHPG, and cortisol.* Psychopharmacol Bull 20:426–430, 1984.

Van Praag HM. *In search of the mode of action of antidepres-*

sants: 5-HTP/tyrosine mixtures in depressions. Neuropharmacology 22:433–440, 1983.

Van Praag HM. *Studies in the mechanism of action of serotonin precursors in depression.* Psychopharmacol Bull 20:559–602, 1984.

Van Praag HM, Lemus C. *Monoamine precursors in the treatment of psychiatric disorders.* In Wurtman RJ, Wurtman JJ (eds). *Food Constituents Affecting Normal and Abnormal Behaviors. Nutrition and the Brain.* Volume 7. New York: Raven Press, pp. 89–138, 1986.

Van Praag HM, Kahn RS, Asnis GM, et al. *Denosologization of biological psychiatry or the specificity of 5-HT disturbances in psychiatric disorders.* J Affective Disord 13:1–8, 1987.

Van Hiele LJ. *L-5-hydroxytryptophan in depression: the first substitution therapy in psychiatry?* Neuropsychobiology 6: 230–240, 1980.

van Vliet IM, Slaap BR, Westenberg HG, Den Boer JA. *Behavioral, neuroendocrine and biochemical effects of different doses of 5-HTP in panic disorder.* Eur Neuropsychopharmacol 6(2):103–110, 1996.

Vanelle JM, et al. *Controlled efficacy study of fluoxetine in dysthymia.* British Journal of Psychiatry 170:345–350, 1997.

Veeninga AT, Westenberg HG. *Serotonergic function and late luteal phase dysphoric disorder.* Psychopharmacology (Berl) 108(1–2):153–158, 1992.

Villaba C, Boyle PA, Claiguri EJ, De Vries GJ. *Effects of the selective serotonin reuptake inhibitor fluoxetine on social behaviors in male and female prairie voles.* Hormones and Behavior 32:184–191, 1997.

Vlasses PH, Rotmensch HH, Swanson BN, Clementi RA, Ferguson RK. *Effect of repeated doses of L-5-hydroxytryptophan and carbidopa on prolactin and aldosterone secretion in man.* J Endocrinol Invest 12(2):87–91, 1989.

Voderholzer U, Hornyak M, Thiel B, et al. *Impact of experimentally induced serotonin deficiency by tryptophan depletion on sleep EEG in healthy subjects.* Neuropsychopharmacology 18(2):112–124, 1998.

Wa TC, Burns NJ, Williams BC, Freestone S, Lee MR. *Blood and urine 5-hydroxytryptophan and 5-hydroxytryptamine levels after administration of two 5-hydroxytryptamine precursors in normal man.* Br J Clin Pharmacol 39(3):327–329, 1995.

Wallace DJ. *The fibromyalgia syndrome.* Annals of Medicine 29:9–21, 1997. Excellent review article.

Wessel K, Hermsdorfer J, Deger K, Herzog T, Huss GP, et al. *Double-blind crossover study with levorotatory form of hydroxy-tryptophan in patients with degenerative cerebellar diseases.* Arch Neurol 52(5):451–455, 1995.

Westenberg HGM, den Boer JA. *Serotonin function in panic disorder: effect of 5-hydroxytryptophan in patients and controls.* Psychopharmacology (Berl) 98(2):283–285, 1989.

Wolfe F, Ross K, Anderson J, Russell IJ, Hebert L. *The prevalence of fibromyalgia in the general population.* Arthritis Rheum 38:19–28, 1995.

Wong DT, Bymaster FP, Engleman EA. *Minireview: Prozac, the first selective serotonin uptake inhibitor and an antidepressant drug: twenty years since its first publication.* Life Sci 57(5):411–441, 1995.

Wurtman RJ, Wurtman JJ. *Brain serotonin, carbohydrate-craving, obesity and depression. In Graziella, Allegri, Filippini, et al. (eds). Recent Advances in Tryptophan Research.* New York: Plenum Press, 1996.

Wyatt RJ, Vaughn T, Galanter M, Kaplan J, Green R. *Behavioral changes of chronic schizophrenic patients given 5-HTP.* Science 177:1124, 1972.

Wyatt RJ, Kaplan J, Vaughn T, et al. *Tolerance and dependence to serotonin : a speculation.* Arch Gen Psychiatry 29:597–599, 1973.

Yaryura-Tobias JA, Bhagavan HN. *L-Tryptophan in obsessive-compulsive disorders.* Am J Psychiatry 134:11–12, 1977.

Yonkers KA, et al. *Symptomatic improvement of premenstrual dysphoric disorder with sertraline treatment.* JAMA 278:983–988, 24 Sept 1997.

Young SN, Gauthier S, Chouinard G, Anderson GM, Purdy WC. *The effect of carbidopa and benserazide on human plasma 5-hydroxytryptophan levels.* J Neural Transm 53(1):83–87, 1982.

Young SN, Pihl RO, Benkelfat C, Palmour R, Ellenbogen M, Lemarquand D. *The effect of low brain serotonin on mood and aggression in humans.* In Graziella, Allegri, Filippini, et al. (eds). *Recent Advances in Tryptophan Research.* New York: Plenum Press, 1996.

Yukitake M, Takashima Y, Kurohara K, Matsui M, Kuroda Y. *Improvement of opthalmoplegia by 5-hydroxytryptophan in two cases of progressive supranuclear palsy.* Rinsho Shindiegaku 36(7):906–908, 1996.

Zmilacher K, Battegay R, Gastpar M. *L-5-hydroxytryptophan alone and in combination with a peripheral decarboxylase inhibitor in the treatment of depression.* Neuropsychobiology 20(1): 28–35, 1988.

ABOUT THE AUTHOR

Ray Sahelian, M.D., obtained a Bachelor of Science degree in nutrition from Drexel University and completed his doctoral training at Thomas Jefferson Medical School, both in Philadelphia. He is certifed by the American Board of Family Practice. A popular and respected physician and medical writer, Dr. Sahelian is internationally recognized as a moderate voice in the evaluation of leading-edge nutrients, herbs, and hormones. In his books and writings, he discusses both the benefits and the risks of these supplements.

Dr. Sahelian has been seen on numerous television programs, including *NBC Today*, CNN newscasts, *NBC Nightly News, CBS This Morning,* and *The Geraldo Rivera Show*; quoted by countless major magazines, such as *Newsweek, Modern Maturity,* and *Health;* medical journals, such as *American Medical News, American Family Physician, Lancet,* and *Annals of Internal Medicine*; and newspapers, including

USA Today, USA Weekend, The Los Angeles Times, The Washington Post, The Miami Herald, and *Le Monde* (France). Millions of listeners, through more than 3,000 radio stations nationwide, have heard him discuss the latest research on health and medical topics.

Dr. Sahelian is editor of *Longevity Research Update,* and the one-million-plus best-selling author of books on melatonin, DHEA, creatine, pregnenolone, glucosamine, saw palmetto, St. John's wort, kava, coenzyme Q_{10}, stevia, and lipoic acid. Many of these books have been translated into several languages, including Japanese, Korean, Italian, German, Russian, and Chinese.

Dr. Sahelian does not sell or endorse products, and has no financial ties to vitamin or pharmaceutical companies. His goal is to interpret the scientific research on nutrition and to provide practical information that can hopefully improve the quality of life of those interested in this type of knowledge. For the latest updates, see his website at www. raysahelian.com.

INDEX